Proceedings of the Institution of Mechanical Engineers

International Conference

Advanced Handling Systems — Applications and Experiences

24–25 May 1988
The Institution of Mechanical Engineers
Birdcage Walk
London

Sponsored by
Manufacturing Engineering Division of the Institution of Mechanical Engineers

Co-sponsored by
Institution of Electrical Engineers
Institution of Mining and Metallurgy
Institution of Production Engineers
British Materials Handling Board
Mechanical Handling Engineers Association
Japan Society of Mechanical Engineers
Verein Deutscher Ingenieure
Société des Ingénieurs et Scientifiques de France

Published for the Institution of Mechanical Engineers by Mechanical Engineering Publications Limited

First Published 1988

This publication is copyright under the Berne Convention and the International Copyright Convention. Apart from any fair dealing for the purpose of private study, research, criticism or review, as permitted under the Copyright Act 1956, no part may be reproduced, stored in a retrieval system or transmitted in any form or by any means, electronic, electrical, chemical, mechanical, photocopying, recording or otherwise, without the prior permission of the copyright owners. Inquiries should be addressed to: The Senior Co-ordinating Editor, Mechanical Engineering Publications Limited, PO Box 24, Northgate Avenue, Bury St Edmunds, Suffolk IP32 6BW.

© The Institution of Mechanical Engineers 1988

ISBN 0 85298 654 8

The Publishers are not responsible for any statement made in this publication. Data, discussions and conclusions developed by authors are for information only and are not intended for use without independent substantiating investigation on the part of potential users.

Printed by Waveney Print Services Ltd, Beccles, Suffolk.

Advanced Handling Systems — Applications and Experiences

Conference Planning Panel

M G Smith, MBE, BSc, CEng, FIMechE (Chairman)
GEC Mechanical Handling Limited
Erith
Kent

G C Bryan, BSc(Eng), CEng, MIEE
Mars Electronics
Wokingham
Berkshire

D Gilson, CEng, FIProdE, FIMatM
Consultant
Upminster
Essex

R H Hollier, MSc, PhD, CEng, MIMechE, MIProdE,
  FIMatM
Department of Management Sciences
University of Manchester Institute of Science and Technology
Manchester

P H Jackson, BSc(Eng)
Mechanical Handling Engineers
London

A Keith, BA, CEng, MIMechE, MIProdE, MBIM
Touche Ross Management Consultants
Guildford
Surrey

P Makin, CEng, MIMechE, MBIM
Health and Safety Executive
Bootle
Merseyside

P Middleton
British Materials Handling Board
Ascot
Berkshire

G F Shimmings
SM Consultants Limited
Windsor
Berkshire

P K Smith, BSc, MSc, CEng, MIMechE
Taylor Hitec Limited
Chorley
Lancashire

J M Williams, BSc, CEng, MIMechE, MIProdE, MRAeS,
  MBIM, SIPDM, FIMatM
Cranfield Institute of Technology
Cranfield
Bedfordhsire

E L Wright, BSc(Eng), CEng, MIMechE
Conveyors International Limited
Leicester

Contents

C92/88	The role of advanced handling systems in running an effective business *P A Dempsey*	1
C71/88	Simple, profitable automatic guided vehicle installations *B D Yates*	5
C72/88	Self-guided vehicle case study proves advanced technology *D F Evans*	13
C73/88	Design of an automatic guided vehicle system *I A Noble*	21
C74/88	Current equipment and new developments *R T Jackson*	27
C75/88	The development of overhead crane robotics for automated handling and storage *C E J Blackstone*	31
C76/88	The application of area gantry robots *P G Hudson*	41
C91/88	Automatic identification — eyes for the computer *M A Marriott*	49
C77/88	The application of automatic speech recognition to parcel sorting and other data-entry tasks *H R Henly*	57
C78/88	Aspects of safety in an advanced handling system *P D Parker*	63
C79/88	Elimination of parcel damage from conveyor systems *C G MacMillan*	71
C80/88	A flexible approach to manufacturing at JCB International Transmissions Limited *I A McDonald*	77
C81/88	Stock handling application in a retail warehouse *A Wright*	83
C82/88	Warehouse control systems *P D Sperring*	95
C83/88	Software implementation procedures for advanced handling systems computer control systems *A St Johnston*	101
C84/88	Integrated control of automated handling systems *M O'Shea*	107
C85/88	Use of intelligent pallets in the Personal System 2 test process *C G McDowell*	113

C86/88	Towards automated assembly using flexible robotic assembly cells, transputers and expert systems *M Taio, R Seals, R Gill, R Ruocco and A S White*	119
C87/88	A comparison of the cost characteristics of modern parts storage and handling systems *K Firth, C Turner and H Paveley*	125
C88/88	The specification of advanced handling systems *R H Hollier and G F Shimmings*	139
C89/88	A case study of the application of computer simulation to automated storage and handling systems *A Keith and K D Porter*	145
C90/88	Project management — the vital ingredient *S J Tomlinson*	149

The Institution of Mechanical Engineers

The primary purpose of the 76,000-member Institution of Mechanical Engineers, formed in 1847, has always been and remains the promotion of standards of excellence in British mechanical engineering and a high level of professional development, competence and conduct among aspiring and practising members. Membership of IMechE is highly regarded by employers, both within the UK and overseas, who recognise that its carefully monitored academic training and responsibility standards are second to none. Indeed they offer incontrovertible evidence of a sound formation and continuing development in career progression.

In pursuit of its aim of attracting suitably qualified youngsters into the profession — in adequate numbers to meet the country's future needs — and of assisting established Chartered Mechanical Engineers to update their knowledge of technological developments — in areas such as CADCAM, robotics and FMS, for example — the IMechE offers a comprehensive range of services and activities. Among these, to name but a few, are symposia, courses, conferences, lectures, competitions, surveys, publications, awards and prizes. A Library containing 150,000 books and periodicals and an Information Service which uses a computer terminal linked to databases in Europe and the USA are among the facilities provided by the Institution.

If you wish to know more about the membership requirements or about the Institution's activities listed above — or have a friend or relative who might be interested — telephone or write to IMechE in the first instance and ask for a copy of our colour 'at a glance' leaflet. This provides fuller details and the contact points — both at the London HQ and IMechE's Bury St Edmunds office — for various aspects of the organisation's operation. Specifically it contains a tear-off slip through which more information on any of the membership grades (Student, Graduate, Associate Member, Member and Fellow) may be obtained.

Corporate members of the Institution are able to use the coveted letters 'CEng, MIMechE' or 'CEng, FIMechE' after their name, designations instantly recognised by, and highly acceptable to, employers in the field of engineering. There is no way other than by membership through which they can be obtained!

C92/88

The role of advanced handling systems in running an effective business

P A DEMPSEY, BSc, DipBA, ACGI, CEng, FIMechE
Rossmore Warwick Limited, Birmingham

This paper places advanced handling systems in the context of the latest thinking on logistics to achieve world class business performance. By case history it questions the need for some systems, emphasises the need for free thinking, simplicity and value, and suggests a methodology to achieve the right system.

In making keynote remarks about Advanced Materials Handling it is my task to touch on the key issues surrounding the subject without necessarily resolving them, and to put the subject into perspective in relation to the businesses to which they are applied. I hope in the next few minutes I can do that with some practical experiences to support my case. The issues about which I shall talk are indeed key issues.

WORLD CLASS COMPETITION AND PERSPECTIVE

Materials handling and storage is a logistical task. In manufacturing business it starts with the manufacture or acquisition of raw materials and finishes with the product in the customers hands backed up by suitable service arrangements.

This pipeline of suppliers, manufacturers and distributors is the full perspective which a conference on advanced handling must address. To concentrate on one element of the pipeline is to limit the overall value to the business. 'Islands of good logistics', have to be linked according to a strategy and planned in just the same way as we link islands of automation and our supply routes to them.

The second important overall perspective of advanced handling is that it covers both physical flows and information flows. Our businesses will not succeed without both elements operating efficiently. Almost by definition this conference will be concerned with physical material flows but we must bear in mind the need to keep simple the demands we are making on information flows.

Materials handling concepts and systems are an essential part of the New Logistics upon which our manufacturing businesses will depend in the next decade. They will provide the essential link between production modules of the future. They will also provide essential links in a much wider array of businesses from mail order distribution to supermarkets and airports.

For example, I am always reminded of the airline commercial which pointed out that in an airport the baggage handling system is the airlines last opportunity to disappoint its customer.

All these types of business are vital to our international competition and image. Handling systems are at the heart of them and the correct choice will decide not only whether a business is competitive or not on a world scale, but may also decide whether a company will exist at all.

THE NEW LOGISTICS

I referred just now to the New Logistics. To a lot of people logistics conjure up a picture of transport issues or perhaps, as with the Falklands War, the complex task of getting fighting personnel and materials to the place where the battle will be fought. Many a battle, and eventually the war, has been lost because the logistical issues went wrong, and not through lack of bravery.

It is a useful picture when we think of businesses that have problems. Whether they recognise those problems or not, they are full of enthusiastic people willing to give their all to keep their livelihood. But the logistical issues can overwhelm them. The New Logistics is the synchronous management of a continuous process of flow of materials, products and information from source to the customer at the right price, quality, quantity, time and cost. Continuous problem solving along this pipeline is vital to:

- eliminate excess and non value adding waste wherever it occurs.
- add value at every step.
- move material directly to where it is needed.
- provide information to everyone instantly (with or without computers).

Many of us have been listening to Hal Mather through the Institutions series of lectures in the past year and are looking forward to hearing him again next year. His ideas on the concepts and philosophies of Just in Time (JIT) are interwoven with the New Logistics management process, but are only a part of it.

The basic Just in Time philosophy is normally applied solely to manufacturing. It can, and must, be equally well applied to all functions of the business and to its supply and distribution lines. I am reminded of a maker of fork lift trucks whose manufacturing engineers conducted a JIT study in his manufacturing plant and was delighted to discover how to reduce manufacturing lead time from 32 days to 6 days. The project did not get investment approval because the investigation also highlighted that the company's order receiving process took about 20 days before the plant received an instruction to build and the distribution process took a further 20 days passing the truck through at least two distribution centres. Since the average customer was prepared to wait only 30 days in total before going to a competitor the actual benefit of the improvement in manufacturing delivery was lost in the inbound and outbound logistics. Changing them involved a vast problem of procedures, traditions, hundreds of managers with their own ideas and interdepartmental squabbles. The change has yet to be achieved at this company.

Nevertheless, these New Logistics involving all functions of the business in both physical and information flow terms will become the culture of successful businesses in products and services in the 1990's. The secret to success will be to have a proper plan for the overall development of your logistics and to stick to it.

DO WE NEED THE SYSTEM AT ALL?

The first essential in establishing a plan is to ask whether you need the system at all, - not good news for the suppliers at this conference but quite often the greatest cost saving question of all.

It may seem an impertinent question to ask but in fact there have been several instances recently where storage and handling systems have been installed and become redundant within a couple of years.

I observed one such example at one of the car assembly plants in Toyota City in Japan. When I walked round the plant they were removing an overhead conveyor which ran the length of the plant from a receiving dock. It carried wheels and tyres. The reason they were removing it was to reduce inventory. They had decided to deliver the wheels and tyres direct to the assembly line. A new hole was being punched through the external wall so that trucks could deliver Just in Time to the point of use.

The conveyor had been in place for just two years. It was 120 yards in length and its removal not only reduced the cost of inventory but saved maintenance cost and reduced noise pollution. When it was installed the planners had not developed their Just in Time delivery thinking to that stage.

I can tell you of another case at Caterpillar in Belgium. This was one of their highest technology and most profitable plants when they applied the New Logistics philosophy as part of their new PWAF (Plant with a Future) plans. The effect was to reduce their work in progress inventories by $14 million in 8 months and eliminate one enormous storage building completely.

SIMPLICITY BEFORE TECHNOLOGY

Now let's talk about the technology and put it in perspective. When we think of Advanced Handling Systems and their importance to our businesses we tend to assume that automation and complexity are inevitable and indeed desirable. This is not the case. It does not follow that advanced handling means automation, high technology, and high capital investment. It does not mean that we have to have computers, AGV's, robots, readers and enormous complexity.

Technology has a vital role but the role of simplicity is greater. It is a human trait, particularly among us engineers to be intrigued by technology. Yet we all know and often forget that the elegance of simplicity in a solution to a problem rules supreme in professional satisfaction.

In a very practical way too, we know that in integrating complex handling systems, the complexity increases in geometric progression with the number of interfaces. A simple solution reduces the number of interfaces and dramatically increases the chances of success and reliability. In other words simplicity is vital to quality.

This is well illustrated in manufacturing by the trend over the last five years to move to cell production. By doing so we have greatly simplified the workflow of a massive variety of components through our manufacturing facilities.

By adopting the approach of simplify, integrate and then automate I have found over the years that frequently 40% of the benefits of a proposed improvement project can be achieved when only about 20% of the planned investment has been made. By taking an independent, free thinking approach to a materials handling task and asking the proper questions, so many costly practices based solely on tradition rather than need become obvious. The mere act of questioning the way things are done often prompts people to make changes and to achieve things they never previously thought possible. This often releases sufficient capital to make the whole project very attractive from the outset.

MAKE IT PAY

Materials handling is a non value adding process. Expenditure on it therefore must be justified by tangible improvement in product cost. The $14 million I have just mentioned at Caterpillar may sound exaggerated to some of you. It is not. Inventory reductions of 60-80% in manufacturing businesses are becoming commonplace when the right principles are applied. Working capital is being released for investment in equipment. Cashflow and profitability are benefitting.

The real benefit of the right handling system is not the initial inventory reduction. It is the ability to sustain that reduction which the handling system provides. In the Caterpillar case this was worth $3.5 million per annum in savings to the cost of carrying inventory.

Nevertheless a recent survey of attitudes to investment shows that direct labour cost savings are still regarded as the main source of justification for capital investment by most medium and small companies. This is wrong, misleading and definitely will not lead to world class performance.

World Class performance will come from considering those investments or changes, whether they are in advanced handling or not, which will most directly benefit the whole business in its market place. By adopting this attitude payback periods, ROI calculations and most other traditional methods of financial assessment take on a much more favourable aspect.

KEEP AN OPEN MIND IN THE CONCEPT STAGES OF A PROJECT TO STAY COMPETITIVE.

The removal of the store at Caterpillar reflected for me a significant change in direction in our whole approach to logistics in manufacturing from which other industries can benefit enormously. After all it was only about six years earlier my colleagues and I had been congratulating a client on completion of installation of an enormous fully computer integrated stores, assembly and test facility for engines. We had helped him concept, design and install the system in record time. The benefits to him at the time and under the current management philosophy had been dramatic - but we would not do it that way now.

By contrast much of the retail distribution industry has not caught up with manufacturing. Recently a major retailing and distribution company was seeking assistance with its inventory and management systems as it forecast inventory increasing from £400m to £600m over the next few years. It also wanted more automated storage warehouses to hold the inventory. A proposal that they should not purchase more storage space and that they should deliver the highest value products direct to retail stores from the manufacturer fell on stony ground.

I do believe that the best manufacturers are now 5 years ahead of retailers in their thinking about warehousing and workflow. The retailing people were still of a storage mentality rather than a product flow or workflow mentality. The situation was not as simple as time permits me to describe here but an open mind would probably have placed them a year ahead of their competition by requiring up to £200 million less working capital and saving the annual cost of carrying the inventory, worth up to £50 million which would go straight to the bottom line.

GETTING THE SYSTEM RIGHT - A METHODOLOGY.

Investment in advanced handling systems has a chequered history of success and failure. In most cases the reasons are clear. It comes down to proper initial planning and attention to detail.

It is not my task to lecture on the subject of methodology but in case you do not have one the following activities at least are fundamental to the manufacturing environment and entirely adaptable to other businesses.

1. Overview and understand the total logistics of the company and plant before embarking on one element of it.

2. Involve people at all stages and explain to them what is happening, how it will affect them and what they can do to help.

3. Simplify the task by analysing the workflows or product flows in terms of their frequency, routing, and value.

4. Design a layout which minimises the amount of handling necessary and gives optimum workflow flexibility and minimum inventory.

5. Decide the minimum necessary information flows and controls to make the systems work. On the whole, ensure they are market driven, not forecast driven.

6. Establish a detailed plan, publish it and stick to it.

7. Write excellent specifications before going to tender. Then buy the most reliable, not necessarily the cheapest and retain ownership of the concept.

8. Integrate each unit carefully, one with another. Automate completely only as the last of several steps in the predetermined plan.

These rules of methodology seem simple and obvious enough yet almost all the failures I have seen, from soft drinks to turbine blades have occured because one of them has been ignored.

SUMMARY

To summarise, there seem to be at least eight key issues in considering development of advanced handling in manufacturing business. I will express them as questions in the context of the New Logistics:-

1. Does the project benefit your whole business (the pipeline) in its worldwide market place?

2. Does it cover both physical flow and information flow issues properly?

3. Do you really need the system at all?

4. Do you have a plan and will you stick to it?

5. Have you simplified the plan before integrating it with the rest of your business and applying the appropriate level of technology?

6. Will the project pay?

7. Have you kept an open mind and tackled traditional attitudes that must change?

8. Have you applied a rigorous methodology to get the system right?

It is my view that these are the discussion issues of this conference which will make advanced handling lead us into the 1990's. These are the issues which will make us world class, and these are some of the issues which will keep us ahead of Europe, as we were last year, in the rebirth and growth of our vital manufacturing base.

C71/88

Simple, profitable automatic guided vehicle installations

B D YATES, MA, MBA, CEng, FBIM
Dexion Limited, Gainsborough, Lincolnshire

SYNOPSIS

AGVs have quick paybacks if the application suits the customer objective. But too many enquirers fail to analyse their true needs and look only at ready made solutions. Some simple small AGV systems are described.

1 INTRODUCTION

For every 45 enquirers wanting an AGV from us only one has an economic case. A prime example of modern technology being wanted "not for dough but for show " [1] Too often systems are requested on the basis of state of the art solutions rather than being requirements driven.

2 AGV STRENGTHS

Why ever should one use an AGV at all? Critics would say they are inflexible, slow, cost the earth to program, need expensive maintenance and cost even more than a fork truck. What could be more useful and flexible than a fork truck with a driver?

Automation can rarely compete with an educated, well managed, highly motivated work force; but how many of us have one of those? [1]

But there are occasions when an AGV comes into its own against conveyors or fork trucks. Let me list some reasons and expand on some of them:

Steady two way material flow with no accumulation

Conveyors are ideal if loads have to be moved quickly and there are peaks and troughs in the flow. A metre length of accumulating pallet conveyor can be had for about £800 but an AGV holding just the same one metre load is going to cost at least £20 000. A conveyor is very much a one way traveller. Hence return loads tend to need another line of conveyor or perhaps a combination of shuttle cars and conveyor.

Control and tracking

The key advantage over a fork lift truck (FLT). This is not an argument for excessive control; indeed we always try to discourage control which does not help achieve the system objectives. But an AGV brings discipline to an environment, cannot be sidetracked by a well meaning progress chaser and does not stop for breaks between destinations.

Hazardous or unpleasant environments

An appropriately protected AGV can travel easily from extreme heat to extreme cold and can also travel in the open air. It can cope with fumes and dust. A machine in Japan has been developed for use in explosive atmospheres [2]. Conveyors in these conditions can rapidly deteriorate. Furthermore AGVs are ideal in clean room environments as they exude less than 1% of the emissions of a man on a truck particularly if urethane wheels are used on a resin floor.

Multishift working

The labour savings in multishift operations are considerable.

Access all round the travel path

Conveyors obstruct floor space and pedestrian access although it is possible to incorporate truck gates in to the heaviest of pallet conveyor. Another conveyor alternative for access is to provide lifting and extending turntables.

Less damage to work in transit

Less spillage of fluid product

Steady acceleration and steady travel speed are a big plus for fragile loads. A malformed pallet going down a conveyor can soon upset a load but that same pallet only bumps once or twice on its journey on and off an AGV. Pallet quality is another area which management ignores at its peril in automated systems but in AGV systems pallets get a lot less knocking about.

Less noise

Less continuous noise indeed but be careful to specify when and why the AGV hooters and flashing lights go off. These signals can be at least as distracting as the steady sounds of conveyor or the occasional FLT going by.

Flexibility of origin and destination

Moving the guide wire is not the problem it used to be with the development of intelligent AGVs with on board microcomputers. Such AGVs also have the ability to do small manoeuvres off the wire (pattern turns) so that the wire does not have to be laid right up to a new deposit station.

3 AGV RELIABILITY

Finally a word about AGV reliability. Look at this table

Table 1 [3]

Reasons for AGV failure

	%
Collisions	41
Electronics	24
Floor problems	14
Leave wire	11
Batteries	7
Emergency stop	3
	100

This paper is not about reliability but the lesson from this table is to think carefully about mixing AGV, FLT and road traffic. The collisions are not with other AGVs but with man driven vehicles!

4 AGV PAYBACK

So the pay back of an AGV system will not be all in direct cash. All the above factors can have a cash value put on them but the accountants may have a different view from the production engineers about the cash values of say improved quality, improved tracking, less product damage, flexibility of overtime costs and flexibility of destination points.

Notwithstanding the difficulties let me illustrate the payback and alternatives in two big examples. The precise locations have been disguised but they are real enough. The first is in a European engine assembly plant. The main option here was a combination of slat conveyor plus overhead chain conveyor. The figures initially looked like this:

Table 2

	System cost DM '000	Savings	Payback Years
Conveyors	3150	255	12.3
AGV	4260	755	5.5

What happened in practice.? Manufacturing cost savings came from a reduction in indirect labour and materials handling. About DM675 000.

Inventory was reduced by DM150 000.

Quality was improved by 9% measured by the number of first time right builds. The AGV system brought discipline to the kitting process and eliminated lineside stocks of small parts. Hence a greater ability of the operators to assemble the engine complete instead of leaving parts off.

System flexibility resulted in one-off later savings. In the past a minor change to the line on new model introduction would have cost DM350 000. Such a change only cost DM180 000 after the AGV system.

This was a large system (78 vehicles) but I use to illustrate the paybacks that can be made in the assembly field. Note that they do not come only from direct cash savings.

Another large (27 vehicles) project in a USA paper warehouse was a difficult payback decision because three major European suppliers quoted prices of $2.1m, $1.4m and $0.9m for the same objective! The payback here based on the lowest quote was estimated at six years.

Table 3 $'000

Costs \ Years	1	2	3	4	5	6	7
Purchase	395	280	275	215	205	105	60
Maintenance	50	70	80	90	90	95	90
Running	255	250	245	245	200	200	195
Total	700	600	600	550	500	400	350
FLT Option	550	550	550	550	550	550	550
Cumulative gain	-150	-200	-250	-200	-50	+150	+++

The user also illustrates the benefit of his AGV system by the graph in figure 1. He feels his real benefits are coming now when he has no need for exceptional maintenance of his AGV system but his FLT experience is that a programme of replacement or heavy maintenance would now be starting.

5 UK ATTITUDES

But what is our attitude to this type of hi-tech investment in the UK? Too many engineering companies set paybacks at two years or internal rates of return over 30%. Contrast that with process industry companies who may set discount rates as low as 8%. But interestingly this latter sector has done consistently better in the U.K. over the last ten years. Distribution businesses seem to fall part way between these two extremes.

Why does UK management set these high discount rates?

Poor research. Directors do not believe forecasts so they filter out the poor projects by making only star certainties look OK

Quarterly results. The stock market will close them down if they don't produce instant profits. So cut costs and stick in traditional niches. In this climate there are no personal returns in long term investment.

Attitudes to risk. To those that do not seek market and development information everything is high risk. Hence (correctly) set a high discount rate to go with the high risk!

Lack of confidence in a long term market. Process companies and retailer/distributors traditionally have relatively secure markets and spend money on marketing. In engineering the winds of competition and technological change can remove markets overnight. Retailing is now becoming more akin to process industries in their long term view of automation.

In Germany they are not so rigorous. In a survey of 15 AGV systems out of 400 known in Germany [4] an interesting statistic was that three companies had not carried out a justification at all. They knew it was right for the long term future of their business. For those who did the payback it was between two and six years. A summary is Figure 2.

6 WORKER ATTITUDES

Just a few words about worker attitudes and company culture. They make or break an installation. An informal fire fighting type of company has a culture that lives on instant reaction to events and a day by day existence. Installation of an AGV system will be a shock. Workers may seek to stop the machines not out of sabotage feelings but to show who is in control. Often the manufacturing side of a company is used to this discipline but the distribution is not. Any project team must look for this culture problem and realise that user training is about the system concept and not just the hardware.

7 SIMPLE PROFITABLE INSTALLATIONS

But this paper is really addressed towards simple profitable installations.

FASSON ADHESIVES

This manufacturer of specialist papers has a work in progress store containing reels of part finished paper weighing four tons. There is a single output/input station at the automatic warehouse and a single point at the factory. One AGV is used to take the paper reels between these two points. The flow is steady at about five reels per hour with a peak rate of nine per hour.

The cost of this system was about £51 000. There is no high level supervisory control. The machine simply goes between stations on requests either from the automatic warehouse or manually from the factory point. The system also allows for expansion in that the circuit can be extended without any supplier software changes; the origin and destination program is written in to the on board microcomputer by the user who can easily add new pickup or deposit stations.

What was the alternative. Conveyors would have cost about £62 000 and taken up valuable floor space. The other option was a fork truck. The factory currently works two shifts and is contemplating a third. Hence at least two drivers with reliefs for holidays would have had to be employed and another driver for the proposed third shift. The driver/FLT unit would have been under employed as there is no other work.

Figures 3 and 4 illustrate this application.

This type of 'fetch and carry' function is AGV automation at its simplest and best. It capitalises on the advantages of AGVs without involving them in high level control systems.

MONSANTO

This application is in the chemical process industry. 24 hour operations mean that any investment involving operators has to take into account that four men have to be hired to run a continuous shift operators job. The layout is in Figure 6.

Why AGV? In the operational area there is much fork truck movement. Hence conveyors would get in the way. The option of overhead conveying was rejected as the necessary lifts would have intruded on manoeuvring space as well as raising the cost.

So what of the FLT option. Firstly the machine would have had to be special anyway as the pallets were up to 4 metres long and could also be a roll cage. But the cost killer was the need to have four men to cover 24 hour working with again low utilisation.

This machine is bi-directional with rollers on its deck running along the machine. Hence the AGV takes pallets on board over its back so to speak, heads out towards the deposit point, drops them off over its front and then returns back along the track in reverse.

The machine is fully automatic being linked into the host computer running the automatic warehouse. This factor plus the two way feature makes it cost about £60 000. But against the option of an FLT the payback is less than one year.

I C L Beehive warehouse, Stevenage

This application is finished goods based and contains five AGVs which are essentially automatic Very Narrow Aisle(VNA) trucks. I include it because it illustrates the lateral thinking that can take place to solve a requirement rather than looking first at a solution. The product did not exist formally until this customer need was highlighted. The early thinking was about existing products such as conveyors, conventional one way and two way AGVs and man driven VNA trucks.

The task is to move pallets of computer hardware from the outfeed conveyor of an automatic warehouse to several floor based picking points or to a remote despatching station. See Figure 7 for the layout.

Conveyors were eliminated early on because of the large number of potential drop off points and the wish to move them regularly. Conventional AGVs were considered but the space available for turning the AGV to drop its load was only 1.8m wide compared with the usual 5m width needed for an AGV to turn in and out of a P & D station.

Hence the look at VNA

Conventional AGVs doing a turn in and out of a P & D station also only go at creep speed for these manoeuvres.

The VNA truck with rotating forks (See Figure 8) can stop in a gap of 1.8m and put its load down to left or right and then proceed straight on. No turning or backing. An ideal solution to this problem. The route is repetitive, the flow regular and tracking of the valuable loads is essential. Solution: a VNA truck with an AGV guidance system interfaced with the main control system.

My main theme has been to look for real paybacks in AGV systems. I have tried to show that although paybacks can come from large systems they are often easier to obtain from small installations of between one and five machines. When presenting this paper I intend to show layouts and photographs of more such simple fetch and carry systems. Indeed we have one in our own factory but part of the payback on this system is that it is a demonstration to our own customers!

A further message is to keep your requirements simple and direct them towards a functional objective. Control where you need to control. Interface with high level hosts only if you need to. Don't be afraid of automatic trucks. Understand their strengths and weaknesses. Look after them and they will give long service without, slowing down with age, stopping to talk to each other and looking for long service medals.

FOOTNOTE

Foreign currencies used in Tables 2 and 3 were worth DM3.00 = £1 $US 1.60 = £1 at the times of the installations.

REFERENCES

[1]   White J A . World class warehousing. 5th International Conference on AGVs I F S (Conferences) Ltd Tokyo 1987

[2]   Kyono S. Inflammable hazardous locations of AGVs. 5th International Conference on AGVs I F S (Conferences) Ltd Tokyo 1987

[3]   Yates B.D. Reliability of Automatic warehouses. 8th International Warehouse Conference. Tokyo 1987

[4]   Daum M Operating costs of AGV systems. 8th International Warehouse Conference. Tokyo 1987

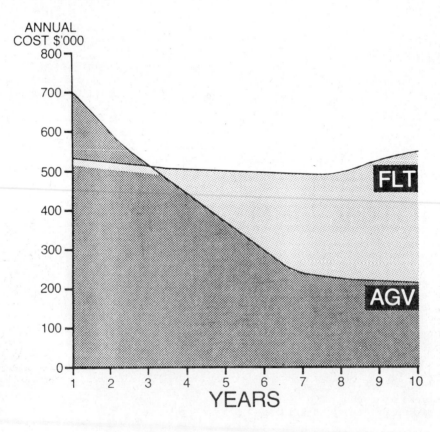

Fig 1   Lifetime cost comparisons — AGV and FLT

AGVs IN GERMANY

System	Payback	Savings	Machines
1	19%	3 trucks	4
2	4 years	Quality	260
3	4 years	People	15
4	4.2 years	People	16
5	4.2 years	People	7
6	None	Trucks	12
7	4 years	10 people	42
8	2.2 years	5 people	1
9	None	20 people	13
10	None	2 people	6
11	28%	Output	5
12	2.5 years	1 million DM	5
13	No data	No data	70
14	6 years	16 men/quality	7
15	4 years	People	8

Source: Fraunhofer Institute Dortmund

Fig 2 Payback of AGVs in Germany

Fig 3 Four-ton load AGV at Fasson

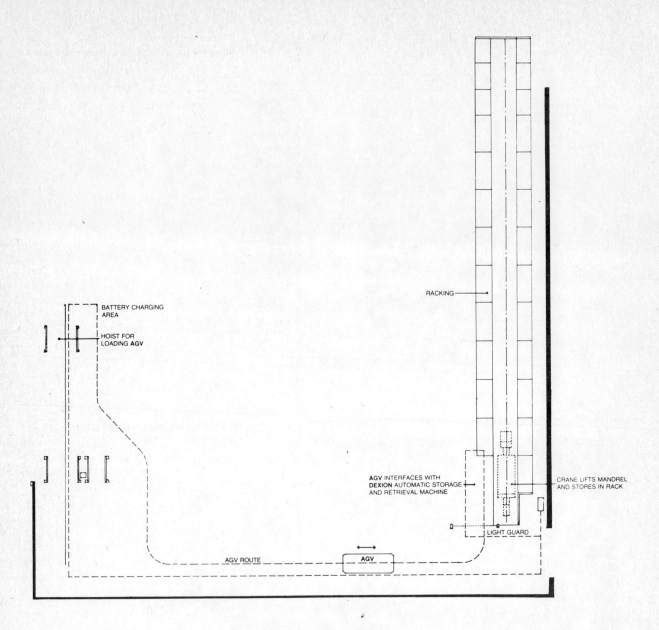

Fig 4 System layout at Fasson

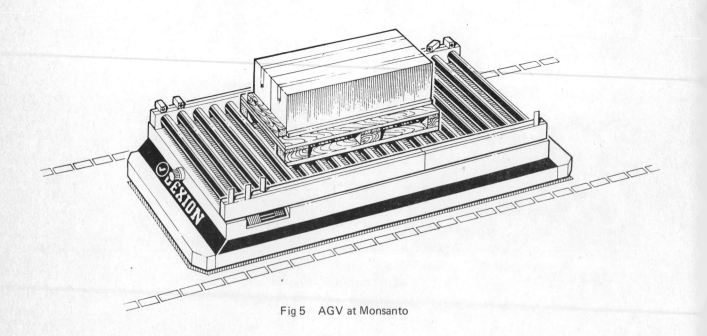

Fig 5 AGV at Monsanto

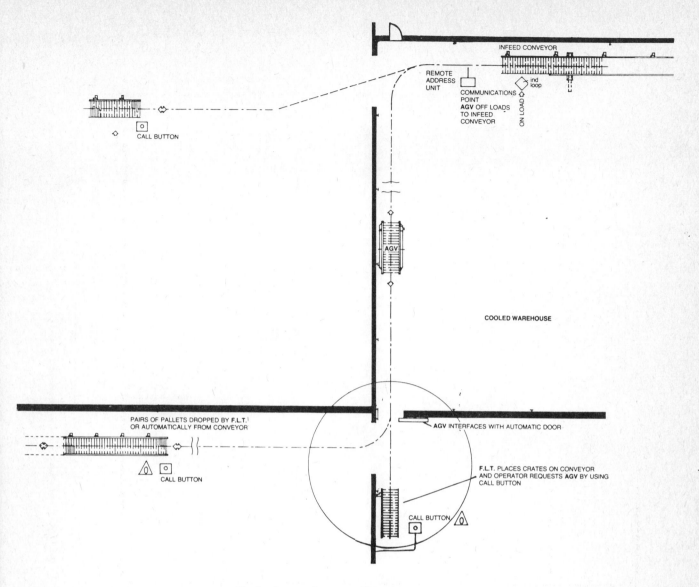

Fig 6 System layout at Monsanto

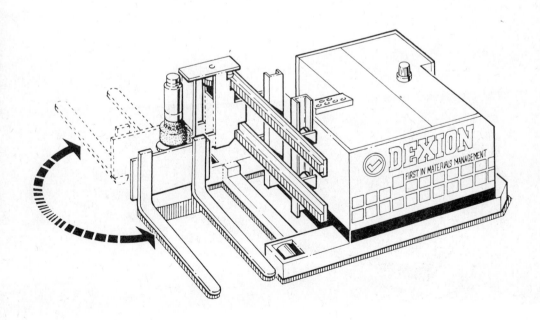

Fig 8 AGV with VNA forks at ICL

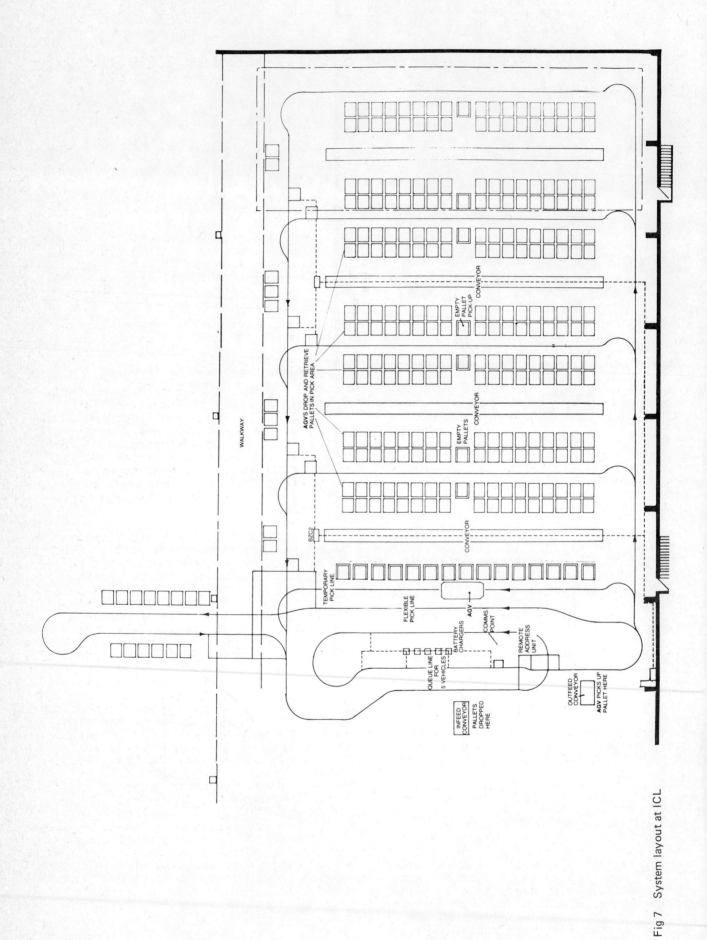

Fig 7 System layout at ICL

C72/88

Self-guided vehicle case study proves advanced technology

D F EVANS
Caterpillar (UK) Limited, Desford, Leicestershire

SYNOPSIS Caterpillar describes how the limitations of conventional AGV's operating in Caterpillar factories were the incentive to develop SGV's - Self Guided Vehicles. Remove the fixed guidepath (wire guidance) and vistas of real flexibility open up for the user. Caterpillar has proved this in its own factories. But the technology of a second generation of guided vehicles is illustrated by an application in the handling system for a tile glazing plant in the UK. No need for special floors. Do it yourself route changes. Operate on same terrain as lift trucks. User friendly system control. True flexibility enhances production capability.

NOTATION

I'm sure by now, everyone here has at least heard of Caterpillar. We've been in the tractor business for over 60 years. Our product line is quite well known, and has the reputation of setting the standard for the earthmoving industry.

Several years ago Caterpillar committed to a complete plant modernisation effort to improve our competitive edge. We decided to modernise our existing facilities, rather than build new ones.

This presents a rather special problem ... continuing production while modernising in buildings with some less than desirable characteristics for the technology on the market. Most of our facilities were built from the 1920's to the 1960's. And some existed before Caterpillar.

AGV's though, obviously provide some flexibility that older material handling systems could not provide. And we have found that AGV's can play an important part in our modernisation plans.

We are not large users of AGV's. However they are critical to some material handling parts of the manufacturing process.

Most AGV's were designed to be installed on smooth concrete floors, of which we have very few. Most of our plants are old, and where we do have concrete floors, they have steel rebar within a few inches of the surface, and steel hardening on the surface. And expansion joints in our factories typically have steel angle iron at every joint. Very few have the smoothness required by wire guided vehicles. Tyre sizes, and the electronics on most wire guided vehicles were not designed to travel on less than SMOOTH concrete floors.

We had two choices. Either buy a few concrete plants and begin capping our floors, or develop a vehicle that can handle these conditions.

We also recognised some of the inherent problems with wire guided technology. Flexibility, although much improved over other automated systems, was limited, because wire-guided AGV paths could not be easily relocated or modified. We saw the need to enable the user to make guidepath additions and changes, in minutes, not weeks or months, and to do it by himself.

Our solution was to remove the physical guidepath. The easiest and most logical way to do this, was to improve the principle of dead reckoning to the point that it could be used as a primary guidance system. However, as those people who have investigated dead reckoning know, the accuracy soon fails as the tyres slide along the floor during turns, and especially on rough conditions such as wood block, expansion joints, manhole covers, and the like.

A process of updating the dead reckoning system is therefore highly recommended. And the most stable part of a factory to reference from, is the factory itself. However the factory must be truly recognisable by the vehicle to be able to track and dock within an acceptable error. By adding 'traffic signs', or barcoded targets, and scanning these with a low power, eye safe laser, we are able to update our dead reckoning system and provide tracking and docking accuracy comparable to existing wire guided systems, while doing it on floors considered unacceptable for wire installation.

The removal of the physical guidepath, creates a new category of AGV's. By replacing the wire guidepath with a coordinate system and giving the vehicle the intelligence to navigate on its own, a Self Guided Vehicle, or SGV, is created.

Vehicle movement is then accomplished by downloading a series of coordinates for aisles and docking points from the Landbase computer, over the radio link, to the vehicles. The navigation is then left to the vehicles to guide themselves to their destination.

The distinguishing feature of a Self Guided Vehicle from a path following system is the Road System concept which permits the vehicle to take the shortest available route between two points instead of being required to follow a loop made by the fixed guide path.

Also by using our fork truck experience, and the associated drive train and electronics as a basis for our SGV design we can now use SGV's just about anywhere a comparable fork truck can be used.

This was a critical element of the vehicle design for us. Simply going 'off wire' wasn't enough. There are many ways to eliminate the wire, but most of them require another type of fixed path. Placing target reflectors along the aisles, on the ceiling, or on the floor, as some have tried, would not accomplish what we wanted if the vehicle is 'following' or 'guiding on' those targets. Path following vehicles are susceptible to 'getting lost' when they lose contact with the path. Rough floors (our biggest problem), other traffic in the aisles, and debris on the floor, can all contribute to the loss of contact with the path. All of these conditions exist in our plants.

The second problem for us with a path following vehicle is that to change the vehicle path, you must move the fixed path. Moving fixed paths takes design and planning time, plus money.

The Caterpillar system stores a co-ordinate system on board the vehicle (in software). Reference point sightings allow the vehicle to update its position dynamically and maintain accuracy. It also provides us with a vehicle which is self correcting for wheel wear and deviations caused by rough floors. Target blockage does not leave the vehicle 'lost' since continuous sighting of targets is not required.

The combination of the navigation system plus a vehicle design that relies heavily upon our lift truck experience, produces a rugged vehicle which can be used just about anywhere normal forklifts are used. Eighteen inch diameter tyres provide an almost all-terrain vehicle. To obtain longer, trouble free operation we used Electronic controls which are ruggedly built and ruggedly mounted. Wherever possible, limit switches were designed out.

However, having good vehicles does not alone constitute an AGV System. The 'system' is just as important as the vehicles, and when more than one vehicle exists, it becomes more important. Computer control is essential for vehicle control, zone blocking and for vehicle optimisation in integrated systems.

And contrary to some reports about this system, large 'main frame' computers are NOT required. The system was originally developed on a WICAT System 150 personal computer. Three of our systems are still running on these PC's. One, is even integrated into a manufacturing system using a DEC 11/73 as a host. All of these systems will eventually be converted to DEC MicroVax II computers for standardisation, but a MicroVax is definitely not a 'main frame'.

Unless you are one of those rare engineers who never changes your factories, the ability to change your guidepath is of great importance. Especially the ability to do it yourself and not in a month or more, but in 15 minutes or less for most changes.

Have you ever tried to change a wire guidepath just 6 inches, or move a docking point an inch? With a Self Guided Vehicle, these changes are a simple do-it-yourself matter. Nothing is changed but a few XY co-ordinates on a very user friendly computer display.

Aisles are simply defined within the co-ordinate system. The desired speed limit is also defined.

Docking points are defined in the same manner, with the coordinate of the desired interface being specified. Cloning of docking points also permits rapid setup of multiple pickup points for deep lane storage.

Zones and schematics are also defined by the user.

This foregoing explains the problems WE had, the theories we evolved towards a solution, and finally the principles that cause our system to work.

And work, it does. The system is no longer a drawing board exercise. It's available.

I can announce that it is in use already, here in the UK. I would like to use this case history to demonstrate to you how the SGV takes materials handling a long stride beyond the AGVs.

H & R Johnson Tiles Ltd, are a large manufacturer - Britain's largest - of ceramic tiles. They are based at Stoke on Trent.

Last year, they decided to automate their production process - including the movement of the materials used in it. To help you understand the SGV's role, let me briefly explain this process.

Wall tiles, varying in size from 6"x 6" to 8"x 10" travel along three automatic lines where glaze is flowed onto them and then loaded - in layers - onto special stillages, either to the kiln for firing, or to the storage area, where buffer stocks are held.

The tiles are fragile and difficult to handle. So the special stillage has been designed to hold them securely.

Each stillage container carries 50 layers of tiles, measures 1530mm wide, 1480mm deep and 3065mm high - including the legs - and has a maximum gross weight of 1880 kilos.

A conventional wire guided vehicle system is inadequate for this application - for a number of reasons.

For example, the size and stability of the load is simply too much. Remember - the overall height is 3065mm and the weight is close to 2000 kilos.

The store - one hundred containers long and two deep - is completely free standing (no locators) so accurate and reliable put-down and pick-up are absolutely essential, the more so when the spacing between stillages is only 100mm.

Our standard Carrier Vehicle is fitted with a special floating lift deck, which has two cups and two cones. These allow the SGV to achieve an exact fit between the stillage and the floating deck.

A microswitch on the deck itself signals the presence of the load. Another microswitch, at the rear of the vehicle, prevents it from overshooting.

To cope with the load size, the rear tactile bumper has been extended to cater for overhang. The front bumper is extended left and right to allow for side overhang. An infra-red scanner at the back of the SGV detects the presence or absence of a load at P & D stations: eliminating the risk of trying to place a load where there's one already, or pick up a load that isn't there!

The sequence of operations goes like this.

When a stillage is filled with newly-glazed tiles, it is picked up by the SGV, responding to a signal from the central computer.

The SGV takes the full container to the buffer store, picks up an empty container and returns to the glazing area.

Accuracy and reliability are essential at the buffer store, when accessing the back row of the store, the SGV can't see any targets - but by relying on its dead reckoning capabilities, it achieves repeatability better than $\pm$ 15mm. The normal spacing between stillages is 100mm.

Responding to another signal from the computer, the SGV will also deliver a full stillage to the kiln.

There are three stillage locations at the kiln loading area - one full, one empty and the third being emptied. As one stillage is emptied, the next is moved into place. Each empty container is returned to the buffer store until required by the glazing line.

Some other statistics.

The SGV uses 24 bar coded targets and receives up to eight requests an hour from each of the three glazing lines and the kiln. The maximum road length is 120 metres. The glazing line operates two eight-hour shifts, five days a week. The kiln works non stop three shifts round the clock - the SGV feeds the kiln from the buffer storage when the glazing lines are idle. By Monday morning when the workers come back the buffer store is empty and the whole process starts again. The system is self-contained - and functions virtually without any human intervention. Application software, written by GEC, manages the store, job priorities, the kiln firing schedule - and gives production information.

In designing the road system full advantage was taken of the ability to alter the routes in the computer software. This allows slow entry into the operations and storage areas, but rapid exit.

The control computer gives an accurate picture of the system status every ten seconds - giving supervisory staff an instant fix on the loading and unloading operations at any time. This information is displayed on the terminal screen.

The land base computer performs a number of key functions.

One is to show the SGV's route map - which is drawn by feeding in aisle, P & D station and target positional information. When the operator needs to make system changes, he can do so via the menu-driven screen and easy-to-use keyboard.

The control operator can also call up other screens:
Stores display - a list of stores and plant locations, showing content and product identification.

The kiln firing schedule.

This identifies which tiles (different sizes and different colours) are required.

A production summary showing the number of stillages which passed through the glazing and firing processes the day before.

Print-outs of the stores display and the daily log can be called up at the end of each day.

The project was implemented in two phases - the first in April 1987, and the second in July.

And I am delighted to say that the company is now looking at putting in a second SGV to serve additional glazing lines, plus a second kiln.

The H & R Johnson application demonstrates a number of things about the SGV system - specific to this case, and also applicable more generally.

Its docking accuracy. Its aisle changing capability. Its stores management. Its total reliability. Its easy integration into the existing productions processes.

This example also confirms the benefits of free ranging SGVs.

No need for special floors.

Do-it-yourself route changes in just a few minutes.

Real time load movement data for inventory control.

A user-friendly control system that eliminates the need to call in experts even for minor changes.

In a word, flexibility.

A flexibility which conventional AGVs, because of the limitations I outlined before, simply don't have - and can't attain because the constraints of their technology, as time has demonstrated, won't allow it.

© Caterpillar (UK) Limited 1988

Fig 1   Tiles travel through glazing weir and are conveyed to the automatic load sequencer

Fig 2 Tiles are automatically loaded, layer by layer with the stillage frame. The frame is located on two cups and cones

Fig 3 When the stillage is full of tiles a message is passed to the landbase control system, which allocates a vacant storage space. The SGV is sent to collect from the end of the glazing line and deliver to the allocated storage space. Note the dusty conditions on the floor

Fig 4 The SGV delivers a load of tiles into an allocated vacant position (shown here stacking two deep). The SGV is working on dead reckoning as it moves between stillages; loads are free standing; no locators in the floor

Fig 5 Free standing loads are accurately positioned with 50 mm clearance either side (photo shows rear looking side sensor). The window in the SGV contains one of the forward obstacle detection infra-red sensors

Fig 6 Responding to a demand from the kiln via the landbase control computer, the SGV picks up a load from the storage area and delivers to the kiln (the positioning of bar code targets can be seen)

Fig 7 The free ranging SGV does not require special floors — route changes are easily accomplished. The truck always takes the shortest available route; all loads are free standing

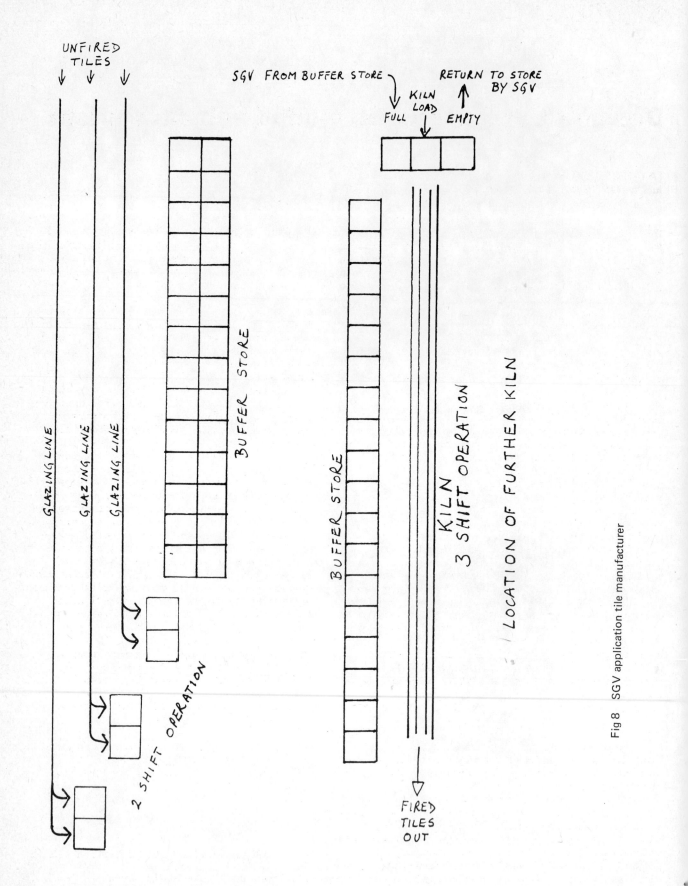

Fig 8 SGV application tile manufacturer

C73/88

Design of an automatic guided vehicle system

I A NOBLE, CEng, MIProdE
Jungheinrich (GB) Limited, Manchester

SYNOPSIS This paper identifies a number of different automatic guidance systems available today. The general areas of application in which automatically guided vehicles are used are described. The main body of the paper is written as a practical guide to help production engineers plan an AGV system which may be used as a method of transporting unit loads around a transport network. All the data which needs to be collected and the factors to be identified are explained.

1 AUTOMATIC GUIDANCE

A number of different methods of automatically guiding the path or route travelled by electrically powered industrial trucks have been developed. The type of truck used as an AGV is generally battery powered, is not attached to 'land' by any rails, wires or cable, runs on industrial concrete floors and can operate in areas used by other vehicles and pedestrians. Up until the last few years such systems have been designed on the basis of specific, pre-designed routes rather like a railway network. Such systems reduce to an absolute minimum the logic required by the vehicle to navigate the route. The majority of operational AGVs installed to date, in excess of 10,000, work on this principle. They have a guidance system equipped to follow a wire embedded approximately 20 mm in the floor and which carries a low voltage AC current. The design section of this paper is based on such a system. Early precursors of this system worked on optical lines painted on the floors or tapes. However, these proved to be unsatisfactory because the lines could be obliterated easily, willingly or otherwise, and in some cases AGVs were distracted by excess sunlight! Surface mounted metal tapes are also vulnerable to damage. Wires which are buried in the floor are almost immune to damage.

Such fixed guidance systems demand a defined transport system and can only work within such a framework. Routes must be kept clear of obstacles and random transport operations outside the scope of the system cannot be catered for. Routes may be changed but this requires physically cutting the floor and laying wire. This regime may be considered beneficial in that a degree of organisation is enforced; random and unauthorised movements cannot take place and material flow takes place according to a pre-defined plan.

'Flexibility' is perceived today to be the way to commercial salvation. Flexibility for AGVs to be able to carry out different tasks and navigate alternative routes at short notice is seen by many to be a desirable goal. The in-floor guidance systems do have a cost which is unique to each application and once installed is relatively fixed inclusive of minor imperfections. In order to overcome this limitation on flexibility, and more important to reduce the overall cost, alternative so called 'free ranging' guidance systems are being developed. The goal of such systems is to be able to emulate, by means of standard software, the ability of the human driver to see obstructions and to be able to navigate the optimum route from point to point; such points to be easily transmitted to and recognised by the vehicle. The computing power to carry out such complex calculations is becoming more readily and cheaply available. The software to carry out such free ranging navigation is expensive to develop, but once perfected may be amortized over the number of vehicles produced. The present development goal is navigation systems to enable the vehicle to recognise its position within its sphere of operations.

The most prominent of such systems uses a scanning laser to recognise strategically placed bar codes. At least four such installations are operational. Other development systems use radio and ultrasonic transmissions.

However for the present the wire guidance system with over 10,000 operational units remains the most reliable and the most practical.

2 AGV VEHICLE TYPES

2.1 Automatic Stacking and Retrieving Trucks

Rider operated narrow aisle warehouse stacking and retrieval trucks which are guided down the centre of aisles by wire guidance are considered by some to be AGVs. This is a very specialist application and should not be considered a true AGV.

A development of this truck is the fully automated stacking truck, primarily designed to automatically stack and retrieve loads in storage racking, but with the extra ability to move to and from the pick up and deposit stations (P & D stations) outside the aisles.

2.2 Flexible Assembly Vehicles (Fig 1)

The AGV used in flexible assembly systems is the basis for many modern assembly plants. The AGV has the ability to transport each assembly module to a variety of assembly stations and remain at each station only for the amount of time necessary to complete the operation. A wide variety of assemblies can thus be processed through the system and the overall time spent in the system is only that required for that particular variation. This method is in sharp contrast to the assembly track in which each assembly has to visit every station and has to remain there for the time of the longest operation. The total assembly time is the summation of all the slowest assembly station times.

AGVs used in assembly can either work on a 'taxi' system in which assemblies are delivered to each station, deposited and collected when ready or the AGV can remain with the assembly and travel through the assembly loop at slow speed with the operators on board. On completion of the operation the AGV is released to travel at full speed to the next operation.

2.3 Tow Tractors (Fig 2)

AGVs are often formulated as a tow tractor which serves a circular route like a bus route. At various points around the route sidings are located in which trailers are loaded and unloaded. When the transport is requested the tow tractor is attached to the trailers, either under driver control or automatically, and takes the train to the next destination. Such a system is ideal for long distance low volume transport.

2.4 AGV with Lifting Forks (Fig 3)

Such machines are often capable of collecting pallets from the floor and delivering to a similar situation. For such an application it is essential that pallets for collection are deposited in the pick up point very accurately as AGVs cannot locate randomly placed loads. This type of AGV has to be able to reverse, which it does at slow speed for safety reasons. Some hybrid AGVs of this type, which normally travel along a wire but have the ability to navigate themselves off the wire at P & D stations, have been developed.

2.5 Unit Load Carrying AGVs (Figs 4 and 5)

AGVs with either simple or multiple roller beds or lift tables are often used to transport unit loads from point to point. Each P & D point has to be equipped with suitable equipment in order to interface with the AGV automatically.

2.6 Order Picking AGVs

AGVs which transport pallets or rollcages through an order picking warehouse enable the operator to order pick continuously without having to drive the truck from point to point. In this application the operator is able to override the automated operation easily at the start and finish of the picking cycle.

2.7 Special Purpose Load Carriers

There are many examples of special purpose AGVs which have been designed to carry paper reels, steel coils, motor cars, petrol and diesel engines, machine tool work plattens and many other specialised products.

3 INFORMATION REQUIRED TO PLAN AN AGV SYSTEM

3.1 Capabilities of an AGV

When designing an AGV system it is important to remember that an AGV does not have the benefit of human sight, feel or intelligence. An ex-factory AGV is useless without a carefully thought out system in which to operate. Vehicles can only collect and deposit loads from very accurately pre-defined locations. Each and every transport operation has to be given to the vehicle as an instruction.

Unless the vehicle is specifically designed to do so, it will not be equipped to detect the presence of a load. Out of place loads, an absence of loads, spilled or damaged loads will not be detected and snarl ups will take place.

This design guide is based on wire guided systems for the transport of unit loads.

In order to plan an AGV system it is essential to define all the factors which affect the operation, to agree the rules of operation (strategy) and to establish the control function.

3.2 Load

AGVs cannot see the load. The maximum and minimum size, shape and weight of the load should be defined. It is unlikely that the same AGV will transport either a pallet or a carton. Standardised unit loads are ideal but not always possible. It is often possible however, to standardise on one pallet, stillage or tote bin size.

3.3 Source and Destination

The source and destination of all loads to be transported should be defined and the throughput established. Both the average and the maximum throughput will be required so the system may be designed to operate with the average volume overall and to cope with individual source maximum requirements.

3.4 Control

The method of task initiation must be decided. The information and the rules governing operation must be established before the method of control can be designed.

3.5 Pick Up and Deposit Station (P & D Stations)

It has been assumed that the loads are situated in a predetermined position. The most suitable type of AGV will determine the P & D required. The P & D may be either passive such as a defined position on the floor, on a static lift table stand or in a rack location. Alternatively it might be interactive such as a roller bed or chain transfer. The pick up locations will usually be equipped with a sensor to detect the presence of the load and to call for service. The interactive deposit station will also have a sensor to enable a signal to be sent to the AGV to indicate the completion of the transfer.

Normal AGVs can stop to $\pm$ 10 mm and P & D stations should be designed to accommodate this. More accurate docking may be necessary for special applications such as machine tools but this will add to the cost and increase cycle times.

3.6 Building Requirements

AGVs always follow the same guide path and therefore stable floors are necessary. Generally tarmacadam floors are unsuitable for this reason. Concrete and sound wood floors are usually acceptable. The floor has to be sound and flat but not necessarily level. Gradients up to 10 per cent can be negotiated. A flat floor is necessary to prevent instability of load and steering, to prevent grounding, to keep all wheels in contact with the floor and to stop vibrations which might damage the vehicle. The floor surface should be kept clean and dry to maintain adhesion for both traction, steering and braking. Gradients should be identified so that descent under full load can be controlled and sufficient power planned to ascend the slope under full load.

Inductive wire guidance can be affected by metal objects in the floor. Large metal objects such as manhole covers and cut off stanchions should be at least 200 mm from the wire. Steel re-inforcement wires should be at least 50 mm below the surface. It has been known for a signal to be generated in a steel re-inforcement bar running parallel to the guide wire and for the vehicle to follow that bar!

AGVs can build up a charge of static electricity. The floor surface should not be insulated as would be the case either with a vinyl covered floor or a floor painted with a non-conductive paint.

3.7 Out of Doors Travel

AGVs can operate out of doors but certain precautions have to be taken. Where possible the routes should be protected from snow and rain. The floor may be electrically heated to melt snow and ice and the route should be freely drained to avoid standing water. Travel from a cold area into a warm humid area will cause condensation which in turn may cause problems with the electrical control of the vehicle.

3.8 Space Required

Safety clearances should be allowed for so that there is always sufficient room on at least one side of the vehicle for a pedestrian to pass; 500 mm is generally accepted as the minimum clearance needed.

AGVs travelling around bends require varying amounts of room, which is dictated by the size of the load, the type of steering and the size of vehicle involved. Lining up accurately with the pick up and deposit points is very important. AGVs should be lined up and travelling in a straight line both on approach to and until the vehicle has cleared a P & D point. It is very difficult to design a system with a P & D station on a bend. Should the area around a P & D point be restricted, it might be necessary to reverse the AGV into the station and leave again along the same path.

3.9 Other Site Interactions

AGVs can operate automatic doors and control other site traffic by operating traffic lights. These junctions have to be clearly thought out so that personnel are not taken by surprise by either the AGV or the operation of the door. If the traffic lights are to be recognised by road vehicle drivers, it is better that they should be of a standard red, amber and green format. The AGV system has to be interlocked with any fire doors to prevent the doors closing whilst the AGV is passing through.

(15cm for Diagram)

4 SYSTEM DESIGN

4.1 AGV Type

The nature of the load to be carried and the method of presentation of the load will dictate the type of load carrier, ie pick-a-back roller bed or chain conveyor, floor level forks, lifting forks, lift table, tow tractor, etc. The choice of vehicle may be modified when the throughput figures have been calculated, because it may be appropriate to carry two or more loads at one time.

4.2 Route Planning

The physical dispositon of all the P & D points and the layout of the building will influence the AGV route. Traffic management is made easier if a one way system using uni-directional AGVs can be devised. It is clear that two AGVs cannot travel in opposite directions on the same wire siumltaneously. In order to balance the traffic flow it will be necessary to designate queue points around the system. There will also be stopping points at which AGVs are inhibited prior to entering a junction, P & D point or any other area which is already occupied by another vehicle. All these points should be planned so that the stationary AGV is clearly visible to other traffic in the area.

4.3 System Dimensioning

Once the routes have been laid out and the traffic volumes to and from each P & D have been established, it is possible to calculate the number of AGVs on the system. The transport system can be considered as a network with each P & D representing a node. The traffic to each node must balance the flow away. Therefore the system must be balanced by the addition of a number of unladen journeys. When the system has been balanced, a matrix can be drawn up on which all the journeys are tabulated. A time value for each journey is established by multiplying the route distance by an average speed. Summation of all the journey times gives the total travel time required. Load transfers require a certain amount of time, typically 30 seconds, so the total time required for load transfers can be calculated. The sum of travel time and load transfer time, usually calculated on an hourly basis, will indicate the total working time required of the system. Should the calculations now indicate that there will be many vehicles using the routes, a factor must be added to allow for the waiting times involved caused by AGVs waiting for each other to clear P & D points and crossings. If only one AGV is to use the route no traffic control waiting time will be necessary.

The service operational requirements now have to considered. If it is more important that the AGV system is 100 per cent utilised, queues will build up and response times to transport requests will become extended. However, should it be important that transport requests are executed as quickly as possible, it will be necessary to build in extra AGV capacity and form AGV queue points around the system.

When the two factors for traffic management and service are added onto the required working time, in minutes per hour, the total is divided by sixty and rounded up thus arriving at the number of AGVs required.

The above calculations form the basis of an AGV system simulation model. The system model can be used not only to calculate the number of AGVs required, but also to assess the effect variables have on the response time of the system.

4.4 System Operational Control

There are several levels of AGV system control, the choice of which is influenced by the method of task initiation and the strategy of task definition.

Task initiation may be originated at three separate levels. The simplest is by manual intervention in which an operator recognises the need for an AGV and summons one by a number of means such as by call button, walking to an AGV in a queue point or by an infra-red torch.

The next level of automation involves a sensor at the P & D station which when activated by the presence of a load releases an AGV from a queue point. Communication to the AGV is often by means of a simple photo cell or infra-red system. When multiple P & D stations are used it may become necessary to queue transport requests in either a programmable logic controller or a small computer.

An AGV system with multiple P & D stations may require a system of priorities other than a queue of requests based on first in first out. In this situation the rules governing the priorities must be carefully considered and defined. A computer is necessary in order to receive such requests and allocate them to AGVs according to the defined strategy. The refinement of this level of control occurs when the AGV control system is fully integrated into an overall factory management system which might change the priorities according to variables outside the transport system.

Task initiation and definition very often occur together but are two different functions. A task definition will be pick up at A proceed to and deposit at B. Such an instruction can be given to an AGV by an operator using a key board on the vehicle. The AGV has to be programmed to recognise such a message and how to execute it. The next level of automation may be operated in a system in which there are relatively few options. Such a strategy might be: When any P & D calls, proceed to that P & D, pick up, proceed to despatch conveyor, deposit and return to queue point. This level of task definition can be handled by a programmable logic controller and simple parallel data transfer to the vehicles. Many transport systems fall into this category for instance: Collection of goods from various production or packing stations and delivery to the warehouse or despatch; delivery of raw materials from a store to various user points.

The next level of task definition which has the ability to collect from any point in a system and deliver to any other point requires much more sophisticated control equipment. The decision of which destination has to be made at the pick up point. This decision can be made by an operator using a key pad, a production machine might define the destination by the nature of the product, another might read a bar code or magnetic code and determine the required destination accordingly. This information then has to be sent to at least a micro-computer which can queue the request together with the details of the task. When an AGV is available this level of data has to be transmitted in serial form via radio or infra-red link to the AGV.

5 SUMMARY OF DESIGN REQUIREMENTS

Collect details of the load size, shape and weight. Determine all the source and destinations for all movements within the system. Quantify as a maximum and as an average on an hourly basis all these movements. Determine the priority strategy for executing the transport tasks. Establish how each task is to be initiated and defined. Plan a route around the area of operation which has enough safety clearance and which minimises the travel distances.

The above information is the minimum which will be necessary to plan an AGV system.

Fig 1 Flexible engine assembly using AGVs as programmable mobile work platforms

Fig 2 A tow tractor AGV towing a train of five hand pallet trucks around a distribution warehouse

Fig 3 A reversing fork AGV for limited lifting operations

Fig 4 One of five roller bed AGVs transferring a load into a bulk store warehouse within a distribution centre

Fig 5 A chain-transfer type of AGV which can also be pedestrian operated

Current equipment and new developments

R T JACKSON, CEng, MIMechE
Consultant, Banbury, Oxfordshire

SYNOPSIS This paper reviews the development of the mechanical handling systems up to the present day, with a brief glimpse of the future. It is intended to put into perspective the subject matter of the Conference, and to stimulate discussion.

1 INTRODUCTION

In introducing this subject, it is first useful to differentiate between Handling Systems as such, and Advanced Handling Systems which have evolved to satisfy the needs of Computer Integrated Manufacture (CIM).

2 CONVENTIONAL HANDLING SYSTEMS

Mechanical handling of Unit Loads has been a fact of life for most of this century, spurred on initially by the vast demands of the North American Market, specifically motor car production.

This demand fuelled the design of simple conveyors for transportation within factories, and sortation and distribution at the many depots situated across the country. Likewise in the motor industry, chain hauled overhead floor and roller conveyors were developed and perfected to produce what we now know as the Production Line.

Almost all of these types of conveyors are still with us today, albeit in a more refined and value-engineered form, and their application has become universal in the developed world.

The advent of electrical, electronic and finally computer control has had major effects on the efficiency and complexity levels achieved by these conveyor systems, whilst leaving the actual conveyor design basically unchanged.

The widespread use of computers in the control of manufacture, and the refinement and automation of the manufacturing machinery has necessitated integrating the handling medium completely into the manufacturing process to form a Computer Integrated Manufacturing System (CIM).

2.1 Conventional handling systems - the hardware

The hardware for these handling methods have changed little over the years.

The main types of equipment are summarised in Table 1 with respect to their main areas of application.

Advances have been made in applications of modern materials such as plastics for small components, and nylons for conveying and driving belts.

Many companies, particularly in the areas of package and pallet handling, have produced whole ranges of standardised conveyors, junctions, turntables, transfer cars and elevators, such that a complete system for handling regular sized loads can usually be assembled from standard component units.

Simple transfer of loads from point to point is often a straightforward case of conveyor selection. Most complications arise at loading and unloading points and transfers between conveyors, and this is where an element of special design is often required.

2.2 Conventional handling systems - the software

Sensing equipment

Control of conventional handling systems has increased in sophistication from the original electrical on-off switch, through interlocked relays, solid-state relays, and memory drum controllers, up to today's systems controlled by Micro-Computers amd Programmable Logic Controllers (PC).

The devices used to sense the progress of the product were originally mechanical limit switches, followed by photo-cells, inductive proximity switches, capacitive proximity switches and even ultrasonic and pneumatic sensors.

With the increasing complexity of handling systems, the number of sensors seems to increase at twice the rate, leading almost inevitably to the use of computers to monitor the high number of inputs thus generated.

With the advent of Programmable Controllers, and their simple programming languages the drawing up of software for handling systems has become a relatively straightforward business, and thus PC's have become almost universally used for direct control of conveying equipment.

All the above devices, with the exception of the photo-cell reflector system are designed to work through computer control, and provide increasing amounts of data-holding capacity, culminating in the solid state transponder, with 100 alpha-numeric or 1K memory available in each small plastic tag.

Table 1 Handling hardware

Load Type	Weight	Handling Type
Packages, cartons and plastic bins	>30kg	Flat/tilt belt or tray sortation Gravity Roller, Ball tables Powered Roller Powered Belt Mini-Stacker Cranes and Shelving Carousel Storage
Steel and timber pallets	>2000kg	Powered Roller Twin Strand Chain AGV Powered Transfer Cars Stacker cranes and racking
Machine pallets, special purpose pallets i.e. engine test cells	>1500kg	Powered Roller Walking Beams Chains AGV Powered Cars Guided Vehicles Stacker cranes and racking
Airport baggage and parcels	>20kg	Close pitched rollers Belt conveyors Inclined Carousels Flat carousels Tilt band sortation Tilt tray sortation
Automobile manufacture		Roller conveyor Chain conveyor AGV Floor or Overhead Chain Conveyor Floor or Overhead Electric Monorail

Identification equipment

Identification of the details of each load or workpiece on a handling system is becoming increasingly important as systems become more complex, and individual workpieces more valuable.

In some systems e.g. distribution warehouses, the correct transportation of loads within the handling system may be checked visually at the end of the process e.g. in the despatch bay. If this visual check has been found to be sufficient, then no further sophistication need be applied.

However, if it is found that more detailed automatic checking is desirable, then a number of techniques may be used as shown on Table 2.

3   ADVANCED HANDLING SYSTEMS

The advent of CIM has generated many new applications for conveyors, and thereby produced radically new types of handling device.

Taking Automatic Guided Vehicles (AGV) as a prime example, they have been around for at least 25 years, and for most of that time have been merely transporting goods from place to place, with manual loading, and more recently loading to and from conveyors.

Today's AGV is still used for transportation from point to point, and as such has taken over the role of the traditional conveyor, especially in cases of low to medium throughput and long or difficult routes. But today's AGV with its inherent flexibility and control, is able to interface directly with the means of production i.e. robot-welders, machining-centres etc. and form part of an integrated production process.

The CIM philosophy of course also uses conventional conveyors, stacker cranes, and rack storage concepts, but the usage of AGV's allows greater flexibility of layout, together with the possibility of using AGVs to collect and deliver tools and swarf containers as well as the work-pieces.

4 NEW DEVELOPMENTS IN HANDLING SYSTEMS

Whilst recognising that conventional handling, storage and distribution of unit loads has now become commonplace, using most of the techniques already described, it has to be said that the biggest advances have taken place where the handling and storage medium have become a part of the automated production process to produce what is termed Computer Integrated Manufacturing systems (CIM).

The correct identification of individual workpieces becomes vital, especially as the process being carried out is usually irreversible, the workpieces themselves are expensive, and also each machine tool is working automatically from a number of available cutting and tooling programmes.

Any error in control or product identification can result in a costly mistake, with possible damage to both workpiece and machine tool, therefore the correct tracking and identification of each workpiece the most important function of the computer management system.

Table 2 Identification devices

Type	Read/Write Method	Information Provided
Adjustable coded reflector	Photocells fixed at a known distance	Simple binary destination code
Bar-coded label, permanent or individually printed	Remote laser scanner / Hand-held scanner / TV recognition	Alpha-numeric code
Magnetic stripe label permanent or individually produced	Hand held reader	Alpha-numeric or digital code
Solid-state transponder, permanent or re-codable	Ultrasonic transmit/receive unit	Alpha-numeric or digital code
Optical character recognition	TV or laser scanner	Alpha-numeric

The impetus for the development of such systems has stemmed largely from changes in the way that products are manufactured e.g. from purpose-built production lines, to a more flexible approach where the tools of production have become more versatile e.g. CNC and DNC machine tools, and the means of transfer between machines of necessity have become integrated. The control of these production systems, or Cells as they are often termed is fully computerised, the necessary local control of handling, storage and machining equipment being provided by individual PC's or Micros built into the equipment.

These Flexible Manufacturing Systems (FMS) are now evolving at a great rate, and usually contain the conventional handling devices mentioned earlier, together with very close control of the workpieces as they traverse the various systems.

4.1 A typical FMS cell

A typical FMS cell would contain the following systems:-

1. Workpiece load/unload stations
2. Work-In-Progress storage
3. Workpiece transportation
4. Production and inspection machinery
5. Workpiece identification
6. Computer system to monitor and control and map Work-In-Progress, and the above systems

These systems have to be assembled from diverse items of hardware e.g. AGV's, conveyors, CNC machines and welding robots. This assemblage of items causes many physical problems at the interfaces between equipment e.g. the positional stopping tolerance of an AGV may be plus or minus 10mm, whereas the workpiece tolerance required by a robot welder may be plus or minus 0.3mm.

However the assemblage of the individual items of software, with the various languages, formats and communication protocols can be the cause of even greater frustration when it comes to writing the control software. Also the Management Software has to be specially written for each application.

The advent of standard communications formats for equipment e.g. Manufacturing Automation Protocol (MAP), should simplify the interface at machinery level, but that still leaves the overall control and production mapping facility as a specially written software function.

5 THE WAY FORWARD

The way forward for automated handling systems will still produce novel methods of moving the products e.g. magnetic levitation, linear motors and air-cushions etc. but the main steps forward are sure to be in the FMS/CIM field with increased use of standardised software packages for the production of goods.

The standardised package would recognise the basic features that are common to all production processes, from the manufacture of steel washers to motor cars, these can be put simply as:-

Receive raw material
Identify and print documentation
Store
Production process
Identify
Store
Production process
Identify
Inspect
Store
Despatch and print documentation

The control of all the conveyors, stacker cranes, machine tools and inspection systems would be based on a similar architecture and the whole system could be integrated with the minimum of problems.

6 THE STATE OF THE ART FMS SYSTEM

A "State of the Art" FMS system would incorporate both the advances being made in faster and higher memory computers, together with the technology which is bringing together the previously separate areas of
Unit Load Handling and Robots.

Using the principles of laser or infra-red communications, together with discrete information transponder units attached to each workpiece, the system will display not only the position of each vehicle, but also the full details of its load status.

The AGV system would be flexible enough to allow an almost infinite variation of routing, based on a map of the area in use, and would thus require a system of guidance which does not rely on fixed floor references other than the load/unload points.

The guidance system of such an AGV is worthy of a separate mention.

6.1 Guidance

Vehicle guidance is by an on-board inertia or laser beam system, which requires no floor wires, cables, markers or magnets. Fine positioning is achieved by an optical alignment device, producing great accuracy.

6.2 Load transfer

The loads are transferred on and off the vehicle with a built in robot arm, which together with the optical alignment device, gives pinpoint loading accuracy. The robot arm is also able to plug-in the on-board battery charger when required, between operations. The load can also be handled with conventional conveyors.

6.3 Communications

Real time two-way communication is maintained by infra-red link, with exact vehicle position, plus the load I.D. being continually monitored.

6.4 Workstation

This unit is the interface between the AGV and the Process Machinery, and is an active station, with real-time control and feedback of all product I.D's being handled.

7 FUTURE TRENDS

The "State of the Art" system as described above is now available from certain manufacturers, and these principles will most certainly extend in future from the FMS/CIM field into general handling and storage applications.

There is no doubt that we shall all be travelling down the FMS/CIM road sooner or later, and I hope that this review of the past and present practice will give you some ideas for your own companies in the near future.

C75/88

The development of overhead crane robotics for automated handling and storage

C E J BLACKSTONE, MA, CEng, MIMechE
Handling Consultants Limited, Stroud, Gloucestershire

Control systems for overhead cranes have been developed to simplify automatic, repetitive, positioning, but to date the mechanical solutions to the vertical storage and retrieval functions of robot cranes have been expensive. The development of a new, simple anti sway reeving system combined with advanced electronic and laser controls changes the picture, with the result that much automated warehousing can be substantially cheaper, while overhead handling in workshops and container depots can be considerably speeded up.

HISTORY

Until the 1940s, few new concepts of handling, as distinct from mechanisation of equipment, had arisen for over 2 000 years. Neolithic man used ropes and by the time the Egyptians started building pyramids, considerable expertise had been developed in the movement and lifting of heavy stone blocks.

By Roman times, hoisting systems were quite common, even to the extent of mobile gantries, while conveyors were developed using wooden rollers. The Industrial Revolution brought about the use of overhead cranes in manufacturing processes in steam driven form, but a Roman legionary would have felt quite at home with the version of shear legs used by the armies of World War II.

The introduction of the fork truck by the US army in the early 1940s completely changed concepts of handling and storage. For the first time unit loads could be moved rapidly round a site with complete flexibility, given the appropriate roads and pasageways, and random access could be achieved in the storage of unit loads by placing these units in a racking structure. The development of the fork and rack principle of handling and storage proceeded apace for the next 20 years. By the 1960s high bay warehouses were being planned which allowed for rapid access to pre-defined unit loads using stacker cranes. Subsequently floor running, high rise, narrow aisle trucks were developed to provide increased flexibility where storage requirement is high in relation to unit flow.

Now the systems design of a store is all important in the definition of the requirement for random access, which is necessarily more expensive than a block stack principle. Considerations of building area available, its cost, the tendency to damage merchandise due to multiple handling, the variations of input and output, and the attendant operational costs in terms of manpower and energy, are all important considerations in the design of the correct system. Nevertheless, there are many finished goods stores and distribution stores where segregation of merchandise in batches is perfectly acceptable, and random access to a part of that batch not essential as long as a practical stock rotation can be maintained. The problem with such a system has generally been that overhead handling is required to maximise storage density and cranes to locate unit loads precisely from an overhead gantry are extremely expensive. Additionally no attachments have been available which enable automatic pick-up of unit loads from a stack while utilising aisles between stacks which are narrower than those through which a man could pass.

The cheapest form of overhead crane is one which utilises a multiple of rope falls to lift the load, but the inherent disadvantage of this in the past has been that the load is free to swing under the crane movement, and this freedom restricts the speed at which the load can be moved. If such a crane is used for block stacking in a warehouse, assistance from an operator working on the floor of the warehouse is invariably required to stabilise and locate the load, or, alternatively, relatively wide aisles are necessary to control the load-carrying attachment suspended from the crab of the crane. In fabrication workshops this weakness slows down movement of work pieces and restricts the speed of assembly where overhead handling is the only practical way of handling large heavy units in maintaining economic utilisation of expensive buildings. Container handling is another area where loading and unloading is restricted by the swinging of the load, while container terminals that can justify stiff masted craneage are limited.

Considerable investment has been made by crane makers throughout the world to overcome the problem of sway from overhead cranes, but the results have invariably been the incorporation of additional hoists providing cross reeving or multiple rope falls from separate drums, which give rise to increased complication and rope wear. A unique, simple system for reeving overhead cranes has now been developed to reduce sway to a minute degree in all directions. The principle of this design is now proven, and the results will revolutionise overhead handling. Warehousing, where batch storage can be accepted, will be substantially cheaper, while overhead handling in workshops and in container depots will be considerably speeded up. Crane attachments have been designed to operate in minimal aisle widths for warehousing operations and crane controls can now be designed to produce a fully automatic warehouse from an overhead gantry crane of basically conventional structure.

TECHNICAL BACKGROUND

In Figure 1, the ratio of 1 on the 'Y' access represents a reeving system that has the same natural stiffness as a pendulum (free swinging forces). It is an interesting phenomenon that for the small rope angles, 2 - 3°, stiffness is no greater than that of a pendulum. The stiffness, of course, depends on the elasticity and loading of the ropes, and there is a limiting value on any angled rope system where the forces would cause one rope to go slack.

The longitudinal elasticity is not constant in a helically wound rope, but decreases at low loads. If the elasticity 'E' is regarded as being constant, then reeving systems where the ropes are lightly loaded statically, (i.e. having a high factor of safety), will experience a smaller stress excursion when resisting natural forces. For a given value of 'E', therefore, the strain excursion will be smaller and it follows that this will produce a stiffer system. Hence reeving systems with low factors of safety on the ropes, and these could be as low as 6 to 1, will be inherently more flexible. If very stiff systems at low rope angles are required, then heavier than normal ropes will have to be used.

The factor of safety on ropes covers two distinct elements. One is rope wear and deterioration and the other is the sundry dynamic loads due to acceleration when lifting or lowering, these latter being usually of very short duration. In a container crane with a high speed crab, the time during which deceleration takes place can be quite significant. If during this deceleration the load is being lowered, then the cables will be passing round the pulleys for some time while experiencing the greater load.

In general, the anti swing reeving system design requires the passage of the rope round more pulleys, and in certain cases with reverse bends. The desire to achieve a stiff system, coupled with the greater rope loads, leads to a choice of static factor of safety greater than normal. This implies the use of heavier cable than normal, but the cable is not usually a significant capital cost item on a lifting system where the anti swing feature is required. Furthermore, with regard to running costs, operational tests have demonstrated that rope wear is no worse than conventional reeving systems for large loads.

A key feature of the anti swing system is that only the main load ropes are used, with no auxiliary ropes or systems. In Figure 2, an arrangement is shown applicable to a monorail hoist which carries a lifting beam that remains level despite off centre loads. This latter feature is provided by the usual arrangement of a twin lift system reeved on to a single drum D. Each lift has two falls of rope, one anchored with the other passing down from the hoist drum; the ropes are numbered on the Figure.

In a single twin lift system the falls RA and RD would be joined round a simple free pulley, hence under normal static and dynamic loading, the tensions in ropes RA and RD would be substantially equal.

In order to provide longitudinal locations for this lifting beam, it must be possible to sustain different tensions in the two falls of the same rope. By extending the rope run to the anti swing device A, in effect a continuous rotary anchor, differential rope forces can be sustained. It is possible to run the falls from the other lift to the same rotary anchor. The rotary anchor does not interfere with the rope movements associated with normal lifting and lowering, while the anti swing facility remains fully effective throughout.

INITIAL DEVELOPMENT

The development of the anti swing reeving system described arose out of a problem of automated handling at the end of a production line for the cutting to length and inspection of high quality finish steel sheets.

These sheets were to be stacked automatically in one of two prime stacking positions, according to grade, in unit masses up to 3 tons. The requirement was to undertake all palletising and special packing away from the production machines, and independent of the existing overhead cranes.

The principle of the layout is shown in block plan, Figure 3, with a perspective drawing of the final installation in Figure 4.

In order that the steel sheets might be stacked without pallets or skids being placed first in the automatic stackers, the stacking is undertaken on scissor lift tables with a rectangular waveform section such that a multi-fork lifting device may be inserted under the stack at alternate equal pitches. These tables, shown as A1 and A2 in Figure 3, travel sideways automatically on completion of a stack, to positions B1 and B2 located under the centre line of a palletising/lifting system suspended from an overhead monorail.
The multi-fork grab suspended from the monorail has features which enable it to palletise a stack of material automatically while it has to be positioned precisely to pick up unstrapped stacks of sheets at positions B1 and B2.

The completed stacks are removed from B1/B2 to positions P for palletising and packing and thence to a transfer position T to remove the completed bundles to positions accessible to the overhead shop cranes.

The maximum mass of completed bundle and the suspended grab could be 5 tons, and the requirement to buffer bundles either at the packing positions or at position X determines that the headroom under the monorail should be some 5 metres.

The development of the anti swing reeving system provided an answer to these requirements at the least cost and complexity.

After some two years operation at two shift working the lifting unit was stripped down for refurbishment of the drive wheels and gearbox. There was found to be minimal wear on the reeving drums and pulleys, while the lifting ropes have only had to be replaced at the same frequency as those for a simple hoist operating under the same duty.

THE APPLICATION TO AUTOMATED WAREHOUSING

High bay Warehousing Systems, with heights in excess of 10 metres to the top of racks, have been in use for some twenty years, but only in the last decade has the software availability and cost brought the overall savings of these systems to the level originally sought.

Manufacturing systems are increasingly being planned on a flexible basis to bring down the unit costs in batch production to close to those previously obtainable only in mass production. These new production systems are ideally linked direct to automated raw material, work in progress, and finished goods storage systems.

The very high bay stores, in excess of 20 metres high, have their place where random access is required to every line item held, notably in spare parts distribution systems, but they are very expensive compared with other systems. For the large proportion of storage applications, perhaps in excess of 90%, such random access is not required and it is perfectly acceptable to visit one of a batch of similar pallets of merchandise.

The height of the fully automated stacker crane installations has been determined in the interests of reducing the capital cost of these stacker cranes in relation to the total number of pallets stored, but problems arise in the inflexibility of use of the building structure, the floor loadings, and planning objections. There is a fundamental advantage in using conventional industrial structures for automated handling, not only in the reduction of the total cost, but in the flexibility to allow the store to be moved, or reorganised, as an adjacent production facility expands, or a distribution requirement is modified. Currently the systems used in conventional industrial buildings utilise floor running trucks, which while only requiring a small amount of floor space, are not suitable for fully automated storage from consideratioons of speed of operation and cost.

Flexible Manufacturing Systems are frequently designed with the use of Automatic Guided Vehicles (AGVs). These, however, take a significant amount of floor space and can be disruptive to other operations. An overhead handling arrangement overcomes these disadvantages, and use of a system minimising the swing of the load overcomes the previous limitations on automated handling from overhead cranes arising from the high cost of suitable units operating from a conventional crane bridge. Thus, similar units can be installed for assistance in the manufacturing process as well as the storage requirements, so eliminating any additional transfer equipment.

The System.

A typical layout of a 'block' store is shown in Figure 5.

A conventional overhead twin bridge crane is located in position on the gantry by means of a rack associated with electronic guidance through a programmable controller. This same system will locate a special form of crab mounted on top of the crane bridges with a configuration which minimises the headroom clearance to the building above the top of the store.

From the crab is suspended a frame containing a close working pallet grab, and this suspension incorporates the patented anti swing system. Figure 6 indicates the arrangement. The use of an overhead crane with what is, in effect, a telescopic stiff legged retrieval unit, facilitates the block stacking of merchandise with cross aisle separations of only 200 mm and 90 mm respectively.

A simple steel structure can be incorporated for stabilising columns of pallets which can be readily moved to an alternative building.

The system for the input and the output of pallets will depend on the nature of the operation, i.e. whether it is full pallet handling or part picking, but the arrangement lends itself readily to input and retrieval on a mezzanine floor with despatch below, providing much more flexibility in this area than conventional automated stacker crane systems.

Comparative Costs.

A cost comparison has been drawn up on the basis of a unit having a gross capacity of 3400 locations. Under average operating conditions the working capacity would be close to 2300 and such a store could be serviced by one Automatic Storage and Retrieval System unit having the ability to travel directly to the desired location.

The aim in this analysis is to compare the alternative types of crane of a design such that their input/output rates are equal for the defined unit of store. In practice, however, additional stacker cranes may well be required to cover the same number of line items, with the result that an overhead handling system may be planned for higher utilisation of the retrieval unit than a stacker system.

Four types of crane operation are compared:

A. Overhead - Anti Swing reeving system.
B. Conventional high bay stacker crane with a transfer unit.
C. Articulated stacker crane, in a high bay, that can drive from one aisle to another.
D. Articulated stacker crane in a conventional building configuration.

The comparative system cost schedule excludes pallets, and is as follows in £ x 1000:

	A	B	C	D
Building Height	9.5m	22m	22m	9.5m
Civil				
Building	475	345	345	490
Site Works	45	45	45	45
	520	390	390	535
Mechanical/Electrical				
Racks	50	135	135	120
Cranes/Transfer	45	155	95	70
Warehouse Computer	25	25	25	25
Input/Output	10	30	30	30
Location Computer	(In A)	35	40	75
	130	380	325	320
TOTAL	650	770	715	855

The important difference between the alternatives which is not shown up in the capital cost analysis is the fact that options A and D use buildings which may be readily used for other functions, whereas options B and C require specialised, inflexible, buildings. The very low comparative figure for equipment in option A offers considerable advantages in terms of low investment coupled with flexibility.

The ratios for building/civil works, against the total capital expenditure schedules are as follows:

	A	B	C	D
Building/Civil = Total	80%	50%	55%	63%

Land costs are not included but in the majority of cases will not be significantly different for the alternative systems.

Depreciation policies are likely to favour solution A, where revised value and general adaptability would justify a longer period for depreciation.

A CASE STUDY

The initial application of the anti swing reeving system has been in the handling of bundles of steel sheet up to 3 tons unit mass, and it is in the area of steel handling in general that the considerable flexibility and adaptability of this approach to automation become particularly manifest.

The materials handling of bundles of steel at the output from strip steel mills is, in general, grossly inefficient in all parts of the world. Finished material is generally held in buildings designed for heavy duty mill operations which are thus both costly and badly utilised in terms of volume and speed of handling. Cumbersome palletising procedures are associated with handling from extravagantly specified cranes or, alternatively, large areas of expensive floor are kept clear for heavy fork trucks. Steel mills are now appreciating the gains to be made in improvement in efficiency with automation in this area of operations and the scope is enormous.

However, the problems of automation, using existing techniques in other industries, are substantial, not only because of the mass of the unit loads, but also because of the variable nature of their size and the packing requirement for different markets. Additionally, the output may also be in the form of steel strip in coil, and the balance between this output and that of bundles can be variable according to market trends.

A case study is now presented, which may be considered typical for this aspect of handling, with examination of the constraints in the systems design, alternative considerations, and a proposed solution.

The constraints.

The existing warehouse at the end of the strip mill is typical, being a vast floor area with many obstructions in the form of control rooms, transformer pens, and sub stations. In addition, there is a rail siding, and access is limited to certain piles by the same equipment. This situation makes the planning of any regularly arranged store impossible unless the floor area is limited. Thus, in order to automate in a racking system, the store must be extended upwards in order to approach the volume required, and this will lead to high ground loading.

The warehouse has to receive bundles of steel sheet, and coils, over 20 shifts a week, while despatch is only served for 10 shifts per week, concentrated in five days. If the handling system worked strictly to this pattern, then, during 10 shifts, the handling rate would be three times that of the remaining shifts, a difference which does not enable key equipment to be effectively deployed.

The requirement for the warehouse is to hold up to 14000 bundles of steel, with unit mass up to 2 tons, and 1200 coils of steel strip with unit mass up to 18 tons. The balance of production could be an increasing percentage of coil at the expense of bundles, and in the case of the bundles, material grain direction and customers unloading and handling facilities, could result in the stillage runners being either way round in relation to the 'length' of the bundle.

The bundles are produced in a wide variety of sizes and each bundle is packed as a single unit with an integral stillage dead size to the cut sheet. Most automatic, or semi automatic handling and storage systems work with a single standard pallet, and it is expensive to produce equipment that will cater for the ranges of sizes involved. Furthermore, if the forks of any handling device are placed at sufficiently close centres to work within the smallest stillage, then the stability and stillage stress is in doubt when handling the largest loads called for.

The existing floor loading is above the design maximum although there are currently no signs of serious settlement or damage. However, a racked storage system will require floor loading capacity some 70% above that at present, and 100% over the original design figures.

The existing warehouse is in full use, and any work needed to strengthen the floor would cause considerable problems. The necessarily lengthy installation of any racking system will prevent the use of the warehouse for a significant period, while the use of a temporary building in the interim poses problems in relation to adequate handling systems.

Systems Considerations.

The cost of automation, whether it be in a warehouse or production facility, can become exhorbitantly high when there is any deviation from standard units for handling and storage, and increased flexibility in a system always costs more than a system with very precisely defined characteristics for operation.

The mill management indicated a requirement for 100% random access to unit bundles, a feature not currently available, as against bundles produced from a specific coil in, say, a batch of five bundles. This can only be achieved with, aside from high equipment costs, acceptance of the following:

(1) Disruption of warehousing and despatch while equipment is installed, with the requirement to provide a temporary building.

(2) Considerable difficulty in holding the required stock in the given area.

(3) Strengthening the existing floor.

(4) Filling in and resiting the siding, or use of part of another bay.

(5) The use of 'slave' boards or standard platens to contain the considerable variations of bundle presentation, with the additional handling involved.

(6) Inflexibility in the predicted change in the balance of output between bundles and coil.

An Automated Stacker Crane System.

The use of automated stacker cranes in association with racking for the bundles and coils will provide 100% random access. However, due to restrictions imposed by plant enclosures and features which cannot be moved, the floor area that a rectangular store can occupy is far less than that presently used for warehousing.

The concentration of loading thus created to retain the required stockholding capacity doubles the floor design load, while a high racking system, with attendant automatic equipment, is intolerant of even small settlements. Thus the existing floor has to be replaced or strengthened.

Racking for the bundles will extend over a total run of 135 metres, in six parallel aisles, with a flow through configuration dictated by the established entry and exit locations to the warehouse. The analysis of the movements indicates the requirement for four high performance stacker cranes. However, as the storage requirement dictates six aisles, even if it were physically possible, the use of crane aisle transfer equipment would be questionable from both operational and cost considerations. A flow through store, with need for access at both ends, precludes the use of such equipment, and therefore six stacker cranes are required.

In order for the variety of unit loads to be handled, it will be necessary for these to be placed on standard 'slave' boards, or alternatively the crane will require automatically adjusted forks. The coils would be held in low racks serviced by a telescopic mast overhead stacker crane which would work in automatic mode within the racks with discharge directly to a despatch stores.

Overhead Automation.

If an overhead robotic system, running on the existing gantry rails and structures, is utilised, the six main disadvantages stated for a racking system are eliminated.

In this instance, the automated handling for bundles of steel would be carried out by two heavy duty overhead twin girder cranes, each equipped for fully automatic operation in all motions, and incorporating a deep reach grab and anti swing reeving system. These cranes can make full use of the existing spare floor area, regardless of existing obstructions, thereby spreading the load as in the current operations.

The capacity and layout of the crane grabs will enable the crane to handle more than one bundle at a time, thus substantially reducing the number of crane cycles per shift, especially during outloading. The anti swing reeving system enables the grab to be kept positioned accurately, vertically below the crab unit. Hence this system enables automatic crane techniques to be applied to wire rope suspended lifting systems.

The capacity of the grab is 7½ tons, while the crane itself will have a capacity of 15 tons in order that an associated cradle unit may be incorporated for yet more rapid handling.

For standard unit loads on 1200 x 1000 pallets, sufficient accuracy for reliable operation may be achieved using standard automatic crane control systems. However, this development poses additional problems of non standard unit loads, wide span rails, and limited fork entry apertures. Therefore a refined version of the system is proposed incorporating established robotic techniques. The grab will be able to sense when a load has been lifted or landed and the forks will be able to 'see' the gap between the stillage deck and the floor or adjacent load. In addition the grab will be able to centre on to the load and read the serial number on an identification tag.

The primary and secondary control systems working in series should achieve an accuracy of: ± 17.5mm across the stillage in the direction of the runners, ± 8mm along the stillage, and ± 5mm vertically. The secondary system takes account of bridge deflection, gantry rail movements and similar problems. The co-ordinates of each load are logged as it is loaded. Each main long travel drive system is separately monitored.

It is considered that the need to extract from the warehouse a particular bundle in a particular sequence should not occur with sufficient regularity to require special provisions. If a shipment needs to be rearranged then this may be carried out in the despatch area. However, since the crane can remove specified bundles then a given bundle can be reached if required. The location of any displaced bundle is, of course, recorded and may be relocated automatically when the crane is free.

If it is considered that the extraction of very small numbers of specified bundles will be a frequent occurence then the crane may be upgraded by adding a bridge cradle. In addition to assisting when selecting a particular bundle, it enables the crane to transport up to six bundles at a time, i.e. three in the grab and three in the cradle.

Coils will be handled in a similar manner using an overhead crane and special grab.
It is a requirement that the facilities shall be capable of accommodating a progressive shift from bundles to coil. To accomplish this, the overhead cranes proposed for handling bundles can be converted to handle coils, as the 7.5 + 7.5 tonne bundles crane will have the same bridge as the 18t coil crane.

An important feature of this system is that the sequence of installation can be conveniently arranged with minimum disruption to the existing storage facility. Since large areas of racking are not involved, most of the floor may remain in use, and only a limited area will have to be allocated to allow installation and commissioning of each crane.

Comparative Pay-Back.

The three principle areas in which cost savings will be achieved by the introduction of automated warehousing systems are:

- Reduction in labour costs.
- Reduction in damage.
- Improvement in performance and service.

The result is a pay-back calculated at $3\frac{1}{2}$ years for overhead automation against $7\frac{2}{3}$ years for automated stacker cranes and racks.

CONCLUSION

The best design is the simple one, and that is the case in this instance with the anti swing reeving system. The sequence of logic to arrive at this solution, however, as is so often the case with simple design, has been a complex one.

The result is that a heavy load suspended from a form of this reeving system will effectively stop 'dead' when the bridge and/or crab from which it is suspended is stopped. Furthermore, considerable forces may be applied to the suspended load in a static position with deflections no more than are experienced with a stiff legged telescopic crane structure. With these characteristics, the way is now open for a complete regeneration of overhead handling systems not only for warehousing, container handling and fabrication shops, but also in the fast moving area of technology in Flexible Manufacturing Systems.

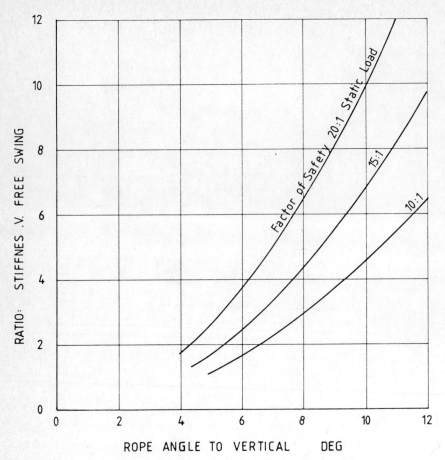

Fig 1 Variation of lateral load stiffness with rope angle and factor of safety

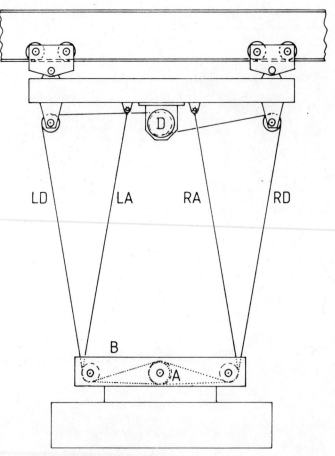

Fig 2 Layout of anti-swing reeving, single twin lift systems

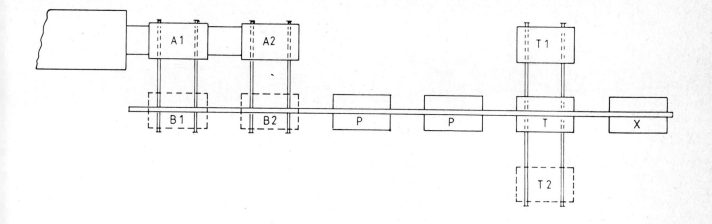

Fig 3 Block layout of handling system for sheet stacks

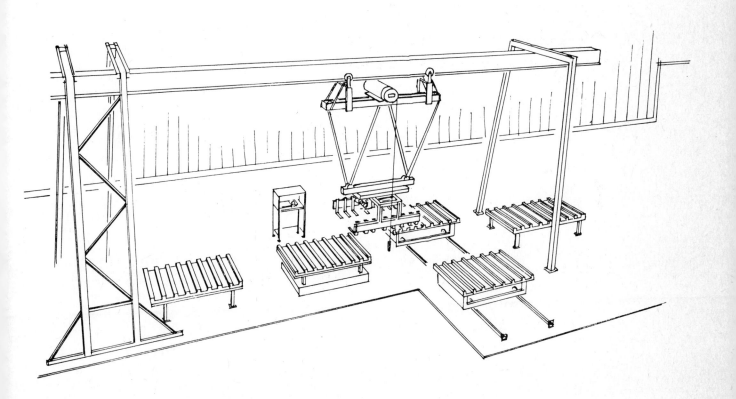

Fig 4 Overall view of handling system for sheet stacks

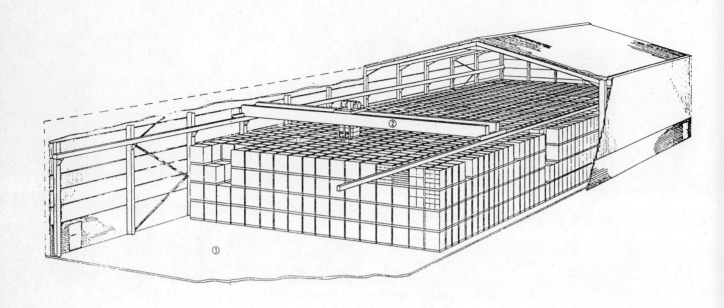

Fig 5 Overall view of a 'block' storage warehouse

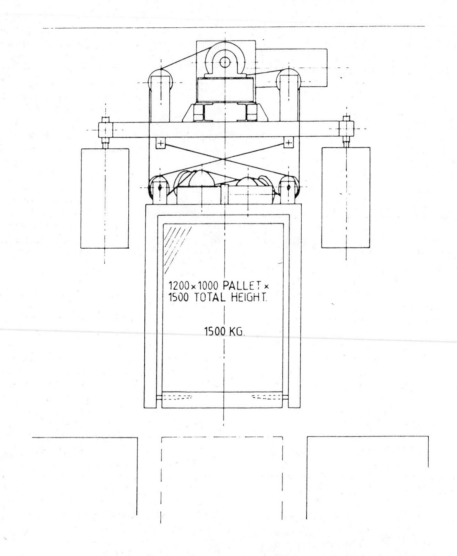

Fig 6 Arrangement of bi-directional reeving system and pallet grab

C76/88

The application of area gantry robots

P G HUDSON, BSc, MPhil
AMECAS, Huddersfield, West Yorkshire

SYNOPSIS This paper describes the application of 5-axes area gantry robots. Seven, 8m x 4m area gantry robots were integrated by the end-user, into a £4.2 million manufacturing system designed to automatically manufacture turbocharger shaft and turbine wheel components from raw material to finished parts. The system consists of 32 successive operations which take place in seven autonomous flexible manufacturing cells. Automatic guided vehicles transport components on pallets between each cell where work is transferred between the pallet and the machines by the 5-axes area gantry robots. To date, it is thought that they are the first and only handling devices of their kind to be installed in the UK.

1 INTRODUCTION

AMECAS, the automation division of Holset Engineering, was created to take advantage of the expertise developed during the installation of the Holset flexible manufacturing system (FMS); a system described by knowledgeable observers as one of the most prestigious in Europe, which was put together in-house in an unorthodox way by 'working upwards from the bottom' rather than the more usual way of starting at the top with sophisticated concepts and then planning downwards.

A team of engineers, required to solve a particular production engineering problem, clearly established the objectives and determined to reach them in the most direct and logical route.

The particular component for which Holset's 1100 square metre FMS was installed at Huddersfield is a shaft and wheel assembly. See fig. 1. This is the single most important component of the turbochargers it makes for heavy duty diesel engines.

It comprises an investment cast Inconel turbine wheel to which is friction welded a forged steel shaft. Revolving at around 150,000 rpm in operation, and with a compressor added on the other end, the turbine assembly uses the fast moving engine exhaust gases to enhance the flow of intake air.

2 THE NEED FOR FLEXIBLE MANUFACTURE

In raw materials alone, the shaft and turbine wheel is the most expensive part of a turbocharger and consequently its production involves a high level of capital tied up in work-in-progress.

The accuracy of machining and quality of surface finish required to enable it to run perfectly balanced and fully floating in oil bearings at such high speeds, make it the most difficult component to manufacture. There are 90 different designs of assembly, between 60 and 125mm turbine diameter for five different frame sizes of turbocharger, built at the rate of 4000-5000 a week.

These factors all work against Holset's objective of being able to supply each customer with the exact design of turbocharger required. Flexible manufacture with its ability to eliminate the traditional problems of small batch manufacture, as well as drastically reduce inventory levels, was seen as the way of overcoming this for at least 50 of the popular sizes of assembly.

The flexibility offered by the new production facility gives design and development engineers much greater freedom in designing turbine assemblies for optimum performance, as well as putting the sales department in a stronger position to offer the customer exactly what he needs.

The installation has reduced the component lead time from 315 hours to less than 24, with turbine assemblies being produced in batch sizes representing the exact customer requirements at the rate of 800 a day, and coming off the line at the rate of one every 1.5 minutes.

The overall project objective was to manufacture shaft and turbine wheel assemblies in small batches at the same unit cost as if they were being manufactured in high volume on dedicated plant. Besides requiring no direct labour, the FMS had to occupy much less space than dedicated plant and achieve zero defects.

Holset's decision to practise end-user project management made it easier to keep within a tight time schedule and control the budget. Standard machine tools were ordered and installed before any of the automation equipment had been finally specified. The machine tools were put into production before their final position or the layout of the facility had been decided upon. This gave an immediate cost reduction.

3 AREA GANTRY ROBOT LOADERS

The automation comprises of 5-axis area gantry robot loaders over each of seven flexible manufacturing cells. A typical gantry robot cell can be seen in fig. 2. Each robot is used to service the machines with workpieces brought to the cell, on pallets, by automatic guided vehicles (AGVs). Under automatic operation the cell gantry loaders are activated by the arrival of a pallet, each with an on-board inductive identification device. The total FMS layout is shown in fig. 3.

3.1 Specifying the machine loading devices

The design of the manufacturing cells was established using a 'bottom-up' approach, starting with the component itself. Factors studied to establish the optimum handling solution included: component type, weight, physical size, range, gripping points, accurate positioning and the pallet design.

Four main configurations of automatic loading device were considered. These were:

a) A single beam gantry with a pneumatic or hydraulic vertical axis and a pallet shuttle device.

b) A multiple axis, floor standing robot.

c) A 3-axis cantilever-type gantry with limited jib traverse.

d) An area gantry robot, with up to six axes, of any physical size dependent upon the area and machines required to be covered.

Option d) was selected because it offered the most flexibility to the project design, an important feature in an FMS installation such as Holset's consisting of many different types of machine processes and consequently a large variety of machines.

Each manufacturing cell was to be individually equipped with an area robot to undertake automatic handling of all work movement within the area, from pallets, through the local processes and back to the pallets again.

Using an area gantry robot to service between two and four machines gave high robot utilisation figures. Taking into consideration the cycle time and physical constraints the design of the manufacturing cells established that up to a maximum of four machines per cell was feasible.

The gantry loaders were ordered early in the project all to a standard 8 by 4 by 2.5 metre high configuration, with a cross-beam mounted vertical ram carrying a double gripper head positioned by two programmable, rotary axes. See fig. 4.

The overall combination provides for five axes of control, which is believed to be unique in the UK., and one of the gantry loaders is in fact a double unit having two independent cross-beam, ram elements operating on the same gantry structure. See fig. 5.

The programmability of the two rotary axes on the double gripper head has proved to be extremely useful when aligning components during automatic loading to machine centres, a problem often associated with single beam loaders. Additionally they are essential to load and unload machines with vertical centres. This flexibility was required in the Holset FMS in those manufacturing cells with a combination of horizontal and vertical centred machines. In fact four of the eight gantry loaders were eventually required to offer components vertically during their operations. See fig. 6.

Two of the installed manufacturing cells feature area gantry robots where each of the six legs have been chocked by 150mm to ensure that the robot head does not foul the machine guarding. This is because the seven gantry structures were ordered to a standard size which assumed the worst possible location for the FMS system due to low services in the roof. This location was in fact the final site for the Holset FMS.

3.2 Interfacing the area gantry robots

The robot controllers were actually ordered 9 months after the loaders themselves; that is not until it had been established to use Heckler & Koch controls nor before Holset had developed its own interfacing specification and instructed all equipment suppliers how the machines should be prepared. All machine tool suppliers were specifically instructed at the time of order that they were not to approach the robot builder for any details. In this way Holset ensured that the project management was kept in-house.

Many machine tool builders when asked to supply an integrated manufacturing cell with material handling, build the whole cell for pass-off at their own works before dismantling it, transporting it to the end-user and then erecting it again. This is expensive and often unnecessary. The first time the robots saw any of the machines at Holset was actually on the shop-floor at Huddersfield.

To ensure this was possible Holset built its own interface simulators, a robot simulator for proving out the machines and a machine simulator for proving the robots. This has meant that all machines have been proved out at the supplier's own works involving the minimum number of visits by suppliers to the Holset factory. Suppliers were simply told how the equipment was to be prepared, what pins on what plugs did what jobs and so on.

3.3 Specifying the detailed gantry robot functions

Because the responsibility for specifying the cell control configuration had been taken by Holset, the detailed gantry robot functions also had be defined.

The following aspects were covered:

a) Area gantry robot M-code specification.
b) Auxiliary features and equipment specification.
c) Machine safety aspects.
d) The various operating modes.
e) Emergency stop procedures.
f) Operator requests to enter machine procedures.
g) Robot to machine electrical interlocking design.

Additional features specified upon the area gantry robots included a component probe on the ram to check for the correct positioning of components and pallets, see fig. 7, auxiliary grippers for automatic component driver change-over, safety shot bolts to physically restrict the gantry and allow personnel to enter the machining areas, special compensating grippers to lock into position when unloading the lathes and wheel grippers unique to the double beam area gantry robot.

Detailed circuit drawings were made of the safety interlocking procedure with the machines and guarding areas. Many electrical devices were added to the guarding to ensure that total safety was achieved. This included the fitting of additional shot bolts and limit switches to every machine guard door, the fitting of infra-red reflective devices above each machine to detect the presence of the gantry robot and the fitting of infra-red curtains across the automatic guided vehicle (AGV) entrance into and out of each gantry cell.

Each M-code function was specified and tested on a robot simulator to ensure the function was correct before application; M-codes were specified for:

a) Automatic programme selection.
b) Signalling presence within pallet areas.
c) Co-ordination of the unload/load sequence with machines.
d) Manual requests to enter the robot working area.
e) Indicating pallet completion.
f) Testing areas for safe entry.
g) Co-ordinating automatic tool change procedures.
h) Controlling main and auxiliary gripper units.
i) Controlling the component probe.
j) Testing for reject components.
k) Actuating safety shot bolts.
l) Issuing fault codes to display panels.

4 A FLEXIBLE, PHASED INVESTMENT

A similar first initiative approach has been applied to the installation of the three AGVs used in the system. See fig. 8. These transport Holset designed plastic pallets, each normally holding between 16 and 32 wheel assemblies, between cells 1 to 7 and a carousel storage area.

When Holset first approached AGV suppliers it found that many would not quote without precise details of the project.

At this stage and for some considerable time afterwards, Holset knew none of these things. In fact the precise site of the FMS was not confirmed until the first machines began to arrive. Moreover, the layout has since been changed and machines re-positioned on many occasions.

Flexible manufacture is about flexible investment. With the AMECAS system, if anybody changes the materials or the process the whole method of manufacture can be re-arranged. Machines can be moved if required, they are all off-the-shelf machines with a standard interface. The line can be easily expanded if required.

With the uncertainty of the final layout, having in-house project management has helped greatly with planning and laying the route of the AGVs. Holset decided for itself which routes the AGVs should take and carried out its own trials with induction wires taped to the floor. As a result, it has finished up with a much tighter route than any AGV supplier recommended.

The cost saving effect of end-user project management and the development of the vital hardware and software has been a factor in enabling Holset to keep within its budget. Of the total £4.2 million project cost, (including a 24 per cent contribution from the Government under its FMS Scheme), £3.7 million has been spent on hardware. The rest was required for the engineering and overheads to complete the system.

5 THE DISTRIBUTED COMPUTER CONTROL NETWORK

Holset recognised that their limitations were in the areas of computer control, interfacing, software development and automated material handling. It was thought to be imperative to acquire this knowledge for in-house use and to work as a catalyst to the educational and training requirements a vital role was played by the local Huddersfield Polytechnic through the Teaching Company Scheme.

It was found that when people were confronted with having to learn new skills and were given the responsibility for the success of the project they responded much more favourably than had they been mere on-lookers.

Although referred to as an FMS it would be more accurate to describe the shaft and wheel assembly installation as seven islands of automation linked physically by an AGV pallet transport system.

Each cell is controlled by a Local Area Control (LAC) consisting of a Heckler & Koch (HK) programmable control and an HK microcomputer. These control the lower level and higher level tasks respectively and are linked directly with the CNC or PLC unit at each machine control. The LAC supervises up to four machines and the gantry robot as well as handling the guarding and safety systems, diagnostics and maintenance systems, the pallet identification system and the AGV communications. Each LAC also talks to the others via a data highway and is connected with a supervisory stores controller. Fig. 9. illustrates a typical cell control configuration.

Cell operations are activated by data picked up from the pallet when it arrives at one of two pallet stations present at each cell. Its inductive, read-write device is an essential part of the data handling system and for production to be started under automatic control the information it carries must correspond with the production schedule.

5.1 The production scheduling

Work schedules are cascaded through the system in such a way that the software pulls work through the cells, beginning at cell 7 and cascading back to cell 1. The production schedule for the shaft and wheel assemblies is determined by the needs of the turbocharger assembly department and it is this demand which is pulling components through the line.

Scheduling of assemblies on the Holset FMS is based on the flow of information in two directions through the data highway linking the local area controllers at each cell. Each cell's demand flows in the opposite direction to the work flow from the finished part stores, down the line, through cell 7 to cell 1 and onto the raw material stores. This has the effect of causing work to be pulled through the system.

Flowing in the same direction as the work flow are the delivery promises calculated by each cell taking into regard their capacity and the delivery promise received by the supplying cell. In addition all pallet monitoring data, cell fault conditions and traffic system data is available at the supervisory stores controller.

This contra-flow of data, which is within 30 seconds of real time, ensures that work in progress and lead times within the system are kept at an optimum.

At the standard level of production the turbine wheel assembly manufacture takes place automatically once components are fed into the beginning of the system at cell 1. From then on, work is carried from cell to cell on AGVs complete with encoded pallets. Each cell processes the parts as received, initiated by data carried within the pallet identification system.

The scheduling system also has facilities for re-routing in the event of non-routine activities in the system. If, for example, a machine within a cell breaks down, partly finished components can be transferred back to the stores by the AGVs continuously patrolling the entire FMS, and taken to manually loaded machines.

6 CONCLUSION

Experience gained through the application of area gantry robots has confirmed that the gantry structure coupled with up to five programmable axes offers a very flexible configuration capable of satisfying most, if not all, of the material handling requirements met in a manufacturing cell.

The most significant constraints to their application include achieving cycle times; especially if a single area gantry robot is used to service several machines with short process times. There is also the physical constraint of roof clearances and the obvious positioning problems when several awkwardly shaped machines and their control cabinets are to be placed underneath a gantry structure.

However, their configuration eliminates the need for pallet positioning tables often required by single axis gantry loaders, eliminates component alignment problems when loading machine tools, saves considerable space that would normally be occupied by floor standing robots and offers the advantage of being able to service positions in both the horizontal and vertical planes. Additionally, high utilisation figures can be achieved when loading several machine tools or processes with the same unit.

The flexibility that area gantry robots offer is useful during project design and installation stages, but most importantly when the installation has been completed and process changes are required because of market changes or product development.

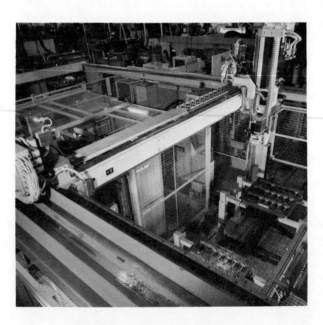

Fig 2 A gantry robot cell

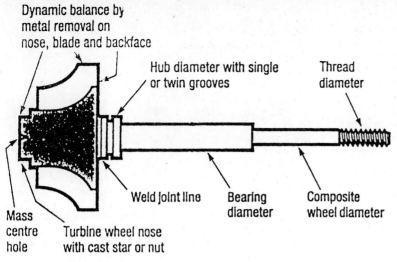

Fig 1 Typical Holset shaft and turbine wheel

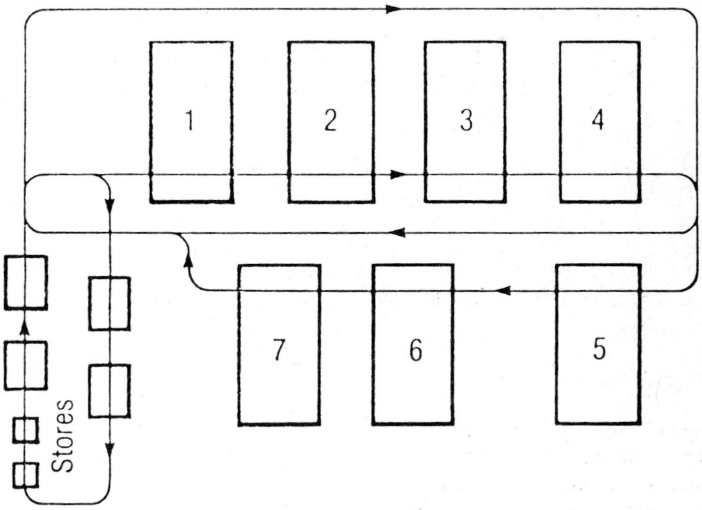

Fig 3 The layout of the seven cells, stores area and AGV routes

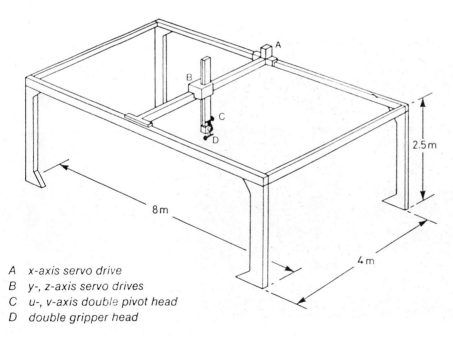

A x-axis servo drive
B y-, z-axis servo drives
C u-, v-axis double pivot head
D double gripper head

Fig 4 Features of a five-axes area gantry robot

Fig 5 Installing a double-beam gantry robot

Fig 7 Probing for component positioning

Fig 6 Loading a vertical-centred machine

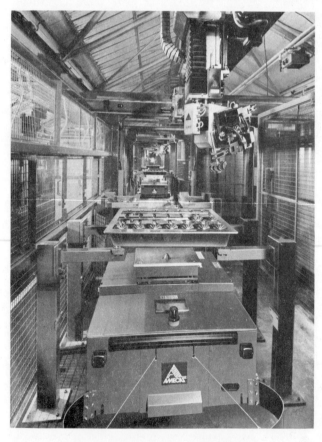

Fig 8 Pallet transportation using AGVs

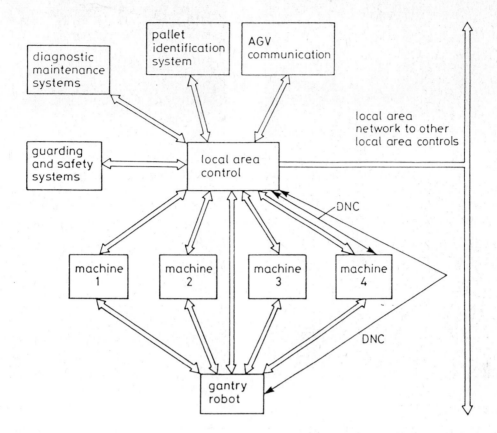

Fig 9 Typical cell control configuration

C91/88

Automatic identification — eyes for the computer

M A MARRIOTT
Numeric Arts Limited, Maidenhead, Berkshire

A review of bar code and other Automatic Identification technologies detailing existing and potential world-wide applications and the importance of technical and industry application standards in maximising the benefits for all.

UNIVERSAL APPLICATIONS

Almost anywhere that computers are used, Automatic Identification is likely to be found. Bar code scanning is used in the London Marathon to record the finishing sequence of participants, whilst fruit packers in Israel use Auto ID to record completed work so that complex piece work rates can be calculated. Even in unlikely places like Papua New Guinea, sophisticated retail scanning systems can occasionally be found. Whether in industry or commerce; government or retailing; New York or New Delhi, Automatic Identification has found universal application.

Auto ID impacts on all our daily lives. Magnetic stripes appear on the plastic cards that we carry in our pocket, and a retail product without a bar code is becoming a rarity.

Today the technology is in use in virtually every sector of the economy:

*   In the electronics industry, bar codes appear on printed circuit boards, components and parts lists.

*   In agriculture, radio frequency tags are now being used to manage herds of cows whilst agricultural chemicals bear the ubiquitous bar code.

*   Radio frequency is used in the mining industry to protect the safety of underground personnel where a switch automatically shuts off conveyors and other machinery if a miner is detected in a danger zone.

*   The construction industry will soon adopt the same technology to monitor the whereabouts of individuals and mobile items.

*   The pharmaceutical industry has used Auto ID for many years and in many applications for example to ensure that drugs, instructions and packaging are correctly matched.

*   In the Blood Transfusion Services, bar codes enable staff to trace the origin of every single donation of blood.

Naturally some applications are more developed than others, whilst an even greater number have not yet even been conceived - in fact Auto ID offers the opportunity for improved profits and efficiency in virtually every sphere of activity.

Let us look at a few established applications:

*   In manufacturing Auto ID is used from goods-receiving through to warehousing and despatch. Factory and office workers 'clock-in' using bar coded or magnetic ID badges. Bar codes are used in piece-part and tool stores as well as on the production line and in quality control.

*   The motor industry is a large user and in the last year has made a major commitment to an international system which will allow goods to move from component manufacturer through to finished assembly using a common Auto ID communication standard.

*   The distribution sector has invested heavily in capital intensive warehousing systems for many years. In some cases, automation depends entirely upon computer memory. Elsewhere however, positive identification is required. Bar code scanning and radio frequency tags are used to monitor the precise whereabouts of pallets as they move into and out of the system. Increasingly, bar coding tracks the distribution cycle from one system to the next. In mail order a bar code is even used to process items returned by the customer.

*   The retail sector has increasingly adopted the technology in recent years. Early systems used Kimball Tags particularly in the fashion sector, and Punched cards were the norm in shoe retailing. At some time or other, almost every Auto ID technology has been tried. Without doubt, in recent years

bar code scanning to the EAN/UPC (European Article Numbering and the North American Universal Product Code) standards has become the retail solution of choice. Fifteen years ago, it was conceived as a system for American supermarkets. Now it is operating in forty countries and is being applied to every type of product.

* Just as you buy goods, so you must pay for them. The banking system was an early user of Auto ID with MICR (Magnetic Ink Character Recognition) on cheques and magnetic stripes on credit cards and cash cards. In some countries OCR (Optical Character Recognition) and bar codes even appear on credit and debit cards.

* Auto ID is commonplace in the publishing world. Many books have bar coded and OCR identifiers. Publishers' catalogues which include bar codes are available to booksellers to speed ordering. If you borrow a book from a library, each copy is likely to be identified with a unique bar code, which can be associated with the borrower through his bar coded membership card. Bar coding is now appearing on some periodicals, and developments are underway which will enable individual articles in scientific journals to be bar coded. Publishers are also experimenting with systems, like Softstrip, to allow data and computer programs to be published in a machine readable form.

* Television and video is a more modern medium. Video rentals are often controlled with bar codes. Where televisions or videos are rented to the customer, the supplier often tracks stocks and repairs with Automatic Identification. Video recorders are now on the market which can be programmed using bar code data entry. Here in the UK this is done from a menu card but in New Zealand, published programme schedules already carry unique bar codes which can be scanned to set the recorder.

* The postal services in many developed countries make use of Auto ID. On letters phosphor dots, OCR and bar codes are all used to aid efficiency and speed sortation. Parcels are often bar coded, as are documents and parcels delivered by couriers. The postal services also use bar codes and radio frequency for sorting mail bags and tote boxes.

* As health care has become more complex and advanced, computers have increasingly been used. Bar codes are now used to identify X-rays, patient files, blood transfusions, drugs and medication, and - in American hospitals - even the patients themselves.

* The use of Auto ID in air transport is already established, and as the pressure on airports and airlines increases, its use will expand accordingly. Some airlines already use bar codes on passenger tickets and for baggage sortation. Scope exists for the use of Auto ID in airline applications ranging from aircraft maintenance management to cargo handling. The most recent development enables passengers and baggage to be precisely matched at the boarding gate, greatly enhancing security.

These are just a few of the countless established applications for Automatic Identification. With new applications emerging daily, the Auto ID industry is continually addressing numerous new challenges posed by eager potential users.

AUTOMATIC IDENTIFICATION : A BREADTH OF TECHNOLOGIES

In today's complex business world, computerised information systems are becoming more and more vital to business efficiency. To achieve this efficiency it is essential that the information going into computers should be accurate, up-to-date and inexpensive to collect. Automatic Identification meets this need.

Auto ID is not a new concept. The idea of machine readable input is nearly one hundred years old, and dates back to the introduction of punched cards. However, it is only since the invention of the computer, and more particularly the micro-processor, the technologies have been developed, refined and have achieved widespread acceptance.

The technologies are of course continually evolving but this does not imply rapid obsolescence. A major market research project was recently undertaken on behalf of AIM-Europe, the industry Trade Association. Its findings indicate that use of existing technologies will continue to expand until the year 2010. Of the four technologies surveyed, 1986 European expenditure breaks down as follows:

	$ m	Share
Bar Coding	787	70%
MICR	171	15%
OCR	138	12%
Radio Frequency	31	3%
	1127	100%

Automatic identification has been defined by AIM as:

> "Technologies which allow information to be coded so that it can be read by machine and processed by a computer so as to minimise human error".

THE TECHNOLOGIES COMPARED

Bar Codes

A bar code is simply an array of parallel rectangular bars and spaces arranged in a particular manner. The arrangement is determined by the encodation rules of the symbology - or bar code language. Different symbologies have the capability of encoding numeric data, alphanumeric data and even full ASCII character sets. Bar code symbols have height solely to provide vertical redundancy, but they are essentially one dimensional images. The height of the bars simply facilitates scanning by requiring less precision.

   Advantages

   * Easily printed
   * Low cost medium
   * Good reading range
   * Good reading speeds
   * Excellent accuracy
   * Low cost readers

   Disadvantages

   * Requires direct vision
   * Symbol space requirement
   * Read only

Bar coding has proved to be a most versatile technology. If the applicaton requires it, millions of impressions of the same bar code can be produced, whilst equally a single unique symbol can be produced on a low cost substrate. Bar code scanners are generally inexpensive whilst at the same time expensive high performance devices are available for demanding applications.

Optical Character Recognition (OCR)

Optical character recognition, allows type fonts to be read by a machine. The range of fonts is limited and for practical purposes this means OCR-A or OCR-B typefaces. The MICR (Magnetic Ink Character Recognition) fonts can also be read optically. Scanning requires two dimensional decoding, which is much more complex than for bar coding. Indeed, to reduce character ambiguity the character set is restricted in some scanners to the numerics plus a limited alpha set.

   Advantages

   * Compact
   * Easily printed
   * Low cost medium
   * Eye readable

   Disadvantages

   * Reading accuracy
   * Reading speed
   * Ease of reading
   * Limited reading range.

In some systems, there are advantages in being able to scan data that the eye can read. A good example is computer generated bills for utilities like electricity, gas and water. OCR is a 'compromise' technology - it must be readable both by machine and human eye.

A different class of OCR scanner - the page reader - has been introduced in recent years. These are fast and potentially versatile but because of cost they are currently used only in mass document handling systems. Transferring such technology to low cost devices is years away.

Magnetic Ink Character Recognition (MICR)

For most purposes, this technology can now be grouped with OCR. Originally, MICR was read by decoding the shape of magnetically encoded characters. The prime application is in banking. Nowadays, OCR scanners are capable of reading these fonts optically.

Magnetic Stripes

These use a similar principle to computer tapes and diskettes. There are two international standards covering this technology. ISO 2894 covers credit cards whilst ISO 3554 covers other applications such as IATA (International Air Transport Authority). Information is held in tracks at densities of up to 210 bpi. Reading is performed magnetically and a stripe can be read 50,000 times before deteriorating. Encoding is expensive, but 'swipe' or badge readers can be very inexpensive.

   Advantages

   * High data density
   * Read and write capability
   * Durable medium
   * Privacy of data

   Disadvantages

   * Cost of media
   * Contact reading
   * Non eye-readable

Magnetic stripes are widely used for bank, credit and cash dispenser cards. It is notable that for many transactions, the magnetic stripe is not read at all, but including the stripe on the card is considered worthwhile for occasions where it is required. Usually, the overall systems which make use of magnetic stripes are of considerable size and sophistication. Applications could boom however with the introduction of electronic funds transfer at the point of sale (EFTPoS). Nevertheless, the technology has been under threat in recent years from 'smart cards' - credit cards with on-board microprocessors.

Radio Frequency Tags

Radio Frequency (RF) technology utilises a microchip contained in a durable outer case. This 'tag' can be programmed with data, which can later be accessed or 'interrogated' via a radio link. The tags can hold as little as 1 bit of data (for presence sensing) whilst the most sophisticated tags hold between 256 and 64,000 bits (for portable databases). Passive tags, like bar codes, OCR and MICR can only be encoded once. Active RF tags, like magnetic stripes, can have their information content updated. Different interrogators can operate from 10mm distance or from up to 1.5 metres. Interrogation can be accomplished through dirt or grease, or through liquids or solid masses. This can be a major advantage.

Advantages

* Environmentally tolerant
* Direct vision unnecessary

Disadvantages

* Cost of medium
* Tag separation essential
* Speed
* Excellent reading range
* High data capacity (Active)
* Read & write capability (Active)

Unlike most other Auto ID media RF tags are too costly to be thought of as 'disposable'. Consequently, at present, RF applications are mainly in closed systems. RF tags can be worn by personnel for access control purposes, whilst mail bag sortation and tracking in postal systems presents another opportunity. In both cases, the RF tags remain under the control of one organisation. International discussions are however taking place which may result in RF tags being used for shipping container identification. This potential "open system" application will however require considerable international co-operation.

INDUSTRY APPLICATION STANDARDS

Open & Closed Systems

All Auto ID technologies can be used in a closed system environment. In such situations, one organisation is responsible for encoding and decoding; once the system is correctly operating, it is likely to remain so.

The technologies are however increasingly being used in open systems. Magnetic stripes are used worldwide in banking and finance. EAN bar coding is now the dominant retail Auto ID technology. Nearly a quarter of a million manufacturers are bar coding products which are being scanned in over 40,000 stores, in 40 different countries. It is plainly impossible for an open system to be successful without the existence of, and adherence to, standards.

Let us now focus on bar coding to examine examples of industry application standards. Whilst the greatest diversity of application and experience relates to bar code, the same principles can of course be applied to all Auto ID technologies.

Symbology Standards

The first fundamental requirement is to have standards for the symbology, this being akin to the bar code 'language'. Surprisingly, all the elements now considered necessary for a symbology standard were only drawn together and published for the first time in 1986. Before then, there were descriptive standards for encoding but details of how to scan, and above all how to decode a symbol, were not placed in the public domain. AIM has now published a set of five Uniform Symbology Specifications covering the following symbologies: Code 39, Interleaved 2 or 5, Codabar, Code 93 and Code 128. There are of course other symbologies still in use today, but effectively these five reflect the industry's view of a rationalised approach to the user's needs.

The AIM Uniform Symbology Specifications include details of:

* code structure
* dimensions and tolerances
* optical specifications

The Uniform Symbology Specifications, for the first time, offer on a comparative basis:

* accurate descriptions of the symbologies
* an explanation of the inter-relationship between printing and scanning
* a reference decode algorithm for the symbology
* a definition of parameters for open systems (industry application standards) and closed systems.

One notable omission from the list of Uniform Symbology Specifications is the European Article Numbering and Uniform Product Code systems. There are a number of reasons for this:

* EAN/UPC is already well documented within an established specification

* Unlike the other symbologies covered by the USS's, EAN/UPC was not developed by a company active under the AIM umbrella.

Nevertheless, the AIM-Europe Technical Committee is liaising with the International Article Numbering Association (EAN) on a number of important technical issues.

Industry Application Standards

An industry standard for bar code or any Auto ID technology, must define certain fundamental features either explicitly or implicitly. Among these are:

* the data to be encoded

* the method of encodation including: symbology, data structure, data length - fixed versus variable, whether data identifiers are to be used and so on

* the symbol size including: the width of the narrow bar, the acceptable range of narrow bar widths, and (if applicable) the width ratio of wide to narrow bars

* the printing constraints including: processes, media, durability, and so on

* the optical specification - not just for scanning but to identify how the specification constrains printing. For example, infra red scanning requires the use of inks with high carbon content

* symbol location and orientation

* scanning parameters.

Whilst these factors must of course be taken into account in a closed system, they are even more vitally important in an open system. This is because often, the organisation responsible for producing the symbol may have no interest, or a different interest, in how the symbol is scanned later on in its life.

By looking at two important application standards we can see the impact of these various factors.

Industry Application Standard : EAN/UPC

Within the EAN and UPC bar code standards a unique 13-digit number is represented in a bar code symbol which has been specially designed for omni-directional scanning, i.e. it is scannable throughout a 360° rotation. Omnidirectional scanning allows staff in retail stores, using flat bed scanners, to have little concern for the orientation of the symbol during the scanning operation. The 13-digit number is structured into four blocks: country of registration, manufacturer code, item code and check digit. This structure confers international uniqueness on any type of product from any manufacturer registered with any one of almost 40 national coding authorities. The symbol has a nominal size dictated by a narrow bar 0.33mm wide. The symbol can however, be printed at sizes ranging between 80% and 200% of the nominal size. This gives manufacturers flexibility to handle small products, and to use a variety of print processes for their packaging.

Because the symbol is incorporated into the product packaging, specific standards were established for precision bar code film master artwork. This film artwork can be incorporated into the package design. The EAN/UPC system was orignally designed around helium neon laser scanners. The scanning beam produced by these devices is at a wavelength of 633 nanometers (red light). This restricts the colour combinations which can be used for bars and spaces. For example, cyan blue is suitable as a dark bar colour, whilst a red bar would be invisible to the scanner. Red would however make a suitable background colour. In effect, all but a few colours which are used to print product packaging can be used to print either the dark bars or the light background of the EAN symbol. A variety of other scanners conform to the original specification, and as a result, over the years, retail scanning has seen the introduction of the widest range of devices in any application.

The EAN/UPC specifications define location, and orientation of the symbol - there are even industry specific standards for books and records - so that scanning efficiency can be optimised. The original scanning parameters required that the symbol had to be scannable from up to 12mm from the scanner. Over the years however scanner performance has increased considerably and symbols can now be scanned from much greater distances, making the whole operation substantially more flexible.

When a manufacturer applies an EAN bar code to his products, he has no certainty as to which retailers in which countries will be scanning that symbol; nor has he any knowledge of the types of scanner which will be used. Furthermore, he has no idea of the developments being considered by scanner manufacturers for the retail market. But he does not need to know any of this. So long as he produces a symbol which meets the requirements of the specification he has played his role in this totally international system. Even if he is scanning the symbol himself, for internal purposes during manufacture or packing, he cannot rely on the fact that it scans on his equipment as proof of acceptability - although this does help. The objective criteria he must meet are those set out in the standards. The closed system acceptance criteria of "if it works it's OK" is no guarantee.

Industry Application Standard : Odette

The next example comes from the automotive industry where an international committee known as Odette, has recently published a standard for a bar coded transport label. The needs of this application are fundamentally different from retail point of sale because of the requirement to apply bar codes to tote boxes, stillages, pallets of raw materials, components and sub-assemblies which will feed into automotive plants throughout Europe.

The first significant difference is that the label requires five different fields of data to be bar coded: despatch note number, part number, quantity, supplier number, and serial number. All this data is encoded in Code 39 symbology. Data identifiers, which are embedded characters to identify the nature of the subsequent information, precede the relevant data. Unlike EAN the data is not of fixed length, nor is it so rigid in structure and format. Fields are nevertheless restricted to a maximum length. The width of the narrow bar can be anywhere between 0.33 and 0.43 mm; whilst the wide to narrow ratio is 3:1. Two spectral bands are specified: B622 in the red light spectrum and B900 in the infra-red. The Odette standards require the symbol to be printed black on white, and to meet the needs of infra-red scanning, carbon inks must be used. The specification requires the five bar code symbols to be stacked vertically above one another, enabling all the symbols to be scanned as quickly as possible using a handheld scanner or automatically using a fixed mount scanner on a conveyor system.

Other Application Standards

These are just two examples. Amongst other bar code standards are those published by:

* The British Footwear Manufacturers - for shoe identification.

* The International Air Transport Authority for passenger ticket identification, passenger baggage handling, cargo handling and cargo container identification.

* The United States Department of Defense - LOGMARS standards for bar coding all procurements by the military. In the UK, the Ministry of Defence has recently established its own bar coding standard.

* The paper industry for newsprint, gravure papers and fine printing papers.

* The health care community in the USA. A similar development is now underway in Europe but this must contend with different requirements in different countries. For example, in Italy non standard "Code 32" bar codes are now printed on most pharmaceutical products, whilst in the UK, pharmaceutical products are soon to be bar coded to the EAN standard.

Industry Application Standards for other Auto ID Technologies

Despite these many examples of bar code standards, some industries have adopted standards for other Auto ID technologies such as:

* OCR for books and banking

* Magnetic stripes for banking, IATA, and elsewhere

* MICR for banking

Newer technologies, like RF on the other hand, have yet to be adopted on an industry-wide basis. The principles and steps necessary for any such implementation will however be similar to those which have applied elsewhere. The first essential step is for a 'demand' to be expressed by the user community.

The Scope for Application Standards

Examples of existing industry application standards make it clear that:

* standards can be applied to a wide variety of industries and applications

* standards can be applied on an international or a national basis

* the requirements of each industry application are sufficiently different to require individual investigation and specific solutions

Although this paper mainly cites examples of industry applications for bar coding, the choice of technology remains diverse. Inspection of a 35mm film will reveal three different technologies in an open system:

* DX coding on the cassette itself - a technology based on electro-mechanical detection

* EAN bar coding on the packaging; on the cassette for laboratory sortation, and as a latent image under the sprocket holes of the film itself

* Raster patterns on the leading edge of the film which can be detected optically or mechanically to identify the characteristics of the film.

The scope and ingenuity of standards is only restricted by the capabilities of the technology. There is potential benefit in standardisation wherever an industry or inter-organisational application can be identified. Industry standards cannot of course address all applications and hence there is still enormous scope for closed or in-house systems.

SOME IMPORTANT QUESTIONS

Let us examine a number of key questions about the relevance of industry standards.

"Why should an organisation or an industry be interested in industry specific standards?"

* Often, different organisations have similar automatic identification problems.

* Sometimes there is an obvious opportunity to 'latch on' to related standards developed elsewhere.

* Sometimes a standard offers benefits which could not be achieved with a series of unrelated closed systems.

In other words, there is a motivation for industry standards wherever it is cheaper or simpler to take this route.

"What should be the objective of any proposed new industry application standard?"

* Because the symbol will move between organisational or location boundaries a standard offers potential benefits in many operations. These boundaries may be the vertical boundaries of the supply chain, or horizontal boundaries as in the retail sector.

* The opportunity exists to create the symbol at source so that wherever it travels in the supply chain, it can be scanned. Often source marking provides an economy of scale which could not otherwise be achieved.

* The symbol can be re-scanned through a series of internal and inter-organisational scanning operations.

* The very existence of an industry standard can result in equipment and systems being developed by suppliers to meet a particular user industry requirement.

"Who develops the standard?"

This is a difficult question to answer. It is however clear from the standards which already exist that:

* there must be demand from the particular 'user' community

* some sponsoring body must provide resources for research, publication and support

* the standard must have input from the user industry, otherwise it will not be adopted

* there should be expert technical input from the automatic identification industry

"How will the system operate?"

This is usually a question entrusted to those preparing the application standard. Some of the issues have been mentioned earlier in the EAN and Odette examples. In summary the issues which have to be faced are:

* What information is to be encoded

* Who is responsible for encoding the information

* How is the encodation to be achieved and physically applied to the item being coded

* How is scanning to be achieved

"Where will the standard be applicable?"

Whatever the start point, it is essential to look up and down the supply chain for scanning opportunities - the more beneficiaries, the greater the chance of the standard being adopted. It is also important to look sideways for commonality with other industries. What they have done or what they are doing might influence the direction of the standard. The question "where?" requires the industry's sphere of activity to be defined, including national and international implications.

The final question is "When?"

This has to do with timescales and phasing, both in terms of development and implementation. Standards take time to develop and it is sometimes difficult for organisations in the vanguard to wait patiently for these developments to take place. If they decide to move forward unilaterally in advance of a standard, they must accept the risk that this entails.

AT THE END OF THE RAINBOW

Generally speaking when standards are implemented there is a need for some threshold level of symbol marking before scanning is economical or beneficial. The past experience of all those waiting and wanting to scan 'standard symbols' generated outside their organisation indicates that it requires persuasion, possibly commercial pressure, and above all patience. Thus, the decision to develop an industry standard is by no means an instant panacea. The development and publication of an appropriate industry standard requires care and commitment. This is usually followed by a significantly longer period of information, education and motivation. The benefits of automatic identification and standardisation are not readily grasped by all. At the end of the day however, an industry standard which has been designed to meet the needs of the entire user community can, without question, offer tremendous benefits to all concerned.

Like nature, Auto ID technology sometimes presents two rainbows. One offering the measurable internal benefits in a closed system. In addition, there may be a second rainbow offering the parallel opportunity to benefit from participating in industry applications standards. In some sectors - retailing is the most obvious example - the 'crock of gold' is only possible with industry standards.

Automatic Identification technolgies have long since proven themselves to be efficient eyes for the computer. With two year payback periods the exception and one year paybacks the norm, it is little wonder that some organisations have identified and implemented dozens of applications for Auto ID.

Automatic Identification is not a technology which can operate in isolation. Its sole purpose is to enhance the operation of other systems. To maximise the benefits of Auto ID it merely requires the eyes of the user to focus on opportunities for improved profits and efficiency through faster, simpler, cheaper and more accurate data acquisition. These are the benefits offered by Automatic Identification.

C77/88

The application of automatic speech recognition to parcel sorting and other data-entry tasks

H R HENLY, CEng, FIERE
British Post Office Research Centre, Dorcan, Swindon, Wiltshire

SYNOPSIS Voice data entry systems are now a viable alternative to keyboards for the input of selection and other data to sorting equipment. This paper describes work which is currently being conducted by the British Post Office to apply this technology to parcel sorting and includes preliminary results from a field trial conducted at a Parcel Sorting Office in 1987.
The application to other tasks associated with the manual handling of mail where hands free operation is desirable is discussed together with some of the problems which have been encountered. The key areas of training and the choice of vocabulary are addressed. Finally the paper discusses the recent developments in the field and points to some of the areas which manufacturers of speech systems should address in addition to simply seeking improved recognition performance.

INTRODUCTION

The Engineering Department of the British Post Office has been actively interested in the application of automatic voice recognition to parcel sorting since 1978. Since that time several different proprietary equipments have been evaluated and two field trials in operational environments have been conducted. The results of the earlier of these two trials together with those of several laboratory trials have been previously reported (Ref.1) and will not be repeated here. The current trial at East London Parcel Office (ELPCO) will be dealt with below. The application of automatic voice recognition to handling tasks such as parcel sorting has the distinct advantage of freeing the operative's hands. In many cases this enables the input task to proceed in parallel with the physical handling task with a consequent improvement in handling rates.

It is abundantly clear from our previous work that the intrinsic performance of the recognition equipment plays only a part in the overall performance of the data input system. The design of the task which incorporates the data input system, the vocabulary used and the work-station design play important roles and still represent areas in which there are problems to be resolved.

1. THE APPLICATION OF VOICE RECOGNITION TO PARCEL SORTING.

The British Post Office operates a parcel network which comprises 23 Parcel Concentration Offices (PCO). The sorting machines in use at these PCOs are, in the main, Tilt-belt machines although there are Tilt-tray machines at Watford, Reading and Glasgow and a Tilt-slat machine at Redhill.

The parcel sorting task is complex in terms of the mental and physical sub-tasks involved.

Several studies have been undertaken aimed at understanding the processes involved in order to improve the productivity of the sorting machine operator and to reduce the level of physical fatigue.
There are two sorting strategies employed on the sorting machines described above. Their use is primarily determined by the work load available at any given time. These are:-
   i). Two man operation. One man selects a parcel from a heap, orients the parcel (ie. faces it) so that the address is uppermost and in a position for reading and places it in a singulated queue feeding the second man. On most tilt-belt installations the provision for this queue is very short and the inter-action between the two men is therefore high. The second man takes a parcel from the queue, reads the address and enters a destination code via a keyboard. He then moves the parcel onto a belt feeding the parcel sorting machine (psm).
   ii). One man operation. In this mode one man performs both of the tasks described above. Currently handling rates for the tilt-belt machine very widely; ranging between 400 to 900 items per hour for one-man operation and up to 1200 items per hour for 2 man operation. In practice the employment of a second man only achieves a 30% improvement in throughput.
In both cases the keyboard may be one of two types depending upon the installation. One type comprises a rectangular array of up to 50 keys, each key corresponding to a unique psm destination chute. The second type is a numeric key-pad comprising ten keys (0-9) and the operator enters the destination number as a two digit string. The keyboard task, because it is combined with the handling task requires a degree of body rotation; a change in the direction of the operator's gaze and therefore eye re-accommodation (focusing) and positioning of the hand on the keyboard to select the required key(s). Neither of the keyboards used are particularly suited to this one-handed operation. In all cases the keying operation interrupts the manual handling task and prevents a rhythmic work flow. The facing and handling tasks involve considerable physical effort; parcels may weigh up to 22 kg and may be large and awkwardly shaped.

Studies conducted in the mid-1970s indicated that at throughput rates of 1000 items/hour with single man operation (ie. facing and coding) between 25% and 35% of the total task time was attributable to the keyboard operation. Amongst the many factors involved motivation plays a large part and sustained handling rates as high as 1600 items per hour have been observed in some offices.

The potential of automatic speech recognition is clear. As a replacement for the keyboard it would enable the data-entry operation to be performed in parallel with the handling task whilst leaving both of the operator's hands free for that task.

Furthermore the performance with a single man would approach that obtained with two men using keyboard entry.

The reduction in body movement and eye re-accommodation are also likely to significantly reduce operator fatigue. However the use of each speech recognition brings its own set of problems and these will be discussed below.

2. SPEECH RECOGNITION SYSTEMS

The last five years have seen significant advances in the range and performance of available recognition systems. Recognition systems fall into two general classes namely 'isolated word' (IW) and 'connected word' (CW) systems. Both types of commercially available systems are currently speaker dependent which means that it is necessary for the user to train the system with his voice and the vocabulary which it is required to recognise before it can be used. Furthermore they can recognise only a restricted size of vocabulary.

The basic recognition process is similar in both classes of machine and uses the principle of the vocoder (VOice CODER). This was based upon extraction of the fundamental frequency of any vowel or voiced consonant and then used a bank of band-pass filters to analyse the remainder of the sound spectrum at any particular time. It was not a speech recogniser but a device for economical voice communication. In a voice recognition system inputted speech is analysed in the frequency domain by a bank of band-pass filters, typically between 10 and 19. A three-dimensional binary pattern is derived which relates the spectral content of the inputted speech with time. This pattern is compared with stored templates which were prepared in a similar way when the machine was trained. The best match is then selected as being the inputted word. In the IW type machine it is necessary that the start and the end of the inputted word is clearly defined and not embedded in a string of words. The CW type machine samples the inputted speech continuously and attempts to determine word boundaries by continuous comparison of the inputted pattern with the stored templates. In both systems time normalisation is used to reduce the effect of varying speaking rates and 'drawl'. Comparison of the inputted speech patterns with the stored templates is done on a statistical basis and the template yielding the highest correlation is selected as the best fit. The algorithm for pattern matching and 'best fit' selection is of course central to the success of the system. Early IW systems supported a vocabulary of only 120 words; current systems provide two or three times this vocabulary size and also the means of extending the size by structuring. Current CW systems have taken the structuring technique further in that they provide the facility for building into the system a simple syntax. The vocabulary size directly affects the time taken to find a pattern match. Currently this varies between 100 and 600 mS for a single word on either type of system and is largely dependant upon the degree to which the members of the vocabulary are discrete ie. not mutually confusable. The variability in this parameter is one of the most serious hardware problems with which we have to cope.

The choice of vocabulary poses several problems. The individual words should be selected so that there is minimum mutual confusion but they must also be selected so as to be relevant to the task and the operator. In the parcel sorting application there are several possibilities. An obvious choice is the use of place-names and/or parcel routings such as Essex, Scotland etc. Some of these comprise two words and would require modification for use with IW systems. Also many of the names currently used are mutually confusable - eg. 'Cork' and 'York', and this mutual confusion varies amongst speakers due to regional accents etc.

Another obvious alternative is the set of numeric destinations used with the PSM keyboard. This produces fewer confusions if the numbers are enunciated as separate digits eg. one-four rater than as fourteen which can be confused with forty! Postcodes could also be used and advantage could be taken of the syntactic structure of CW systems to ensure that a code was of valid form. However the alphabet contains several phonetically confusable letters and it would be necessary to employ a phonetic alphabet eg. the ICAO/NATO code - to eliminate this source of confusion. Experimental work conducted for us by UWIST has indicated that, owing to the smaller vocabulary size, the increase in processing time would not be excessive. Furthermore, since a structured alpha-numeric code would be required to replace place-names there would be scope for recovery from errors which a place-name system does not provide.

In designing a vocabulary to minimise confusion it must be appreciated that words which sound different to the human ear do not appear that way to a recognition system. The human ear is much more adept at distinguishing between short consonantal sounds than is the speech recogniser which relies more heavily on distinguishing the longer vowel utterances.

The larger vocabularies pose other problems also. Training of the machine, that is establishing the recognition templates for each user, is time consuming and error prone where it requires several repetitions of each item. Also, even with relatively small vocabularies the user has difficulty in learning and remembering the vocabulary words (cf.Ref 2). Furthermore, it is difficult to ensure that the user trains the machine with the same 'voice' as the one he will use in practice. These effects have been observed in our experimental work at East London Parcel Office.

3. ERGONOMIC AND OTHER FACTORS WHICH AFFECT PERFORMANCE.

The environment in which the system is used is very important particularly in respect of ambient noise. The currently available speaker-dependent recognition systems are only 'noise' processing machines! Any inputted noise can be given an interpretation and if it recurs then it will be recognised. Likewise if the machine receives noise it will attempt to recognise it. Noise coinciding with inputted information will corrupt that information by altering its spectral content, resulting in failure to recognise the input. Experiments conducted by the German Post Office have shown that continuous noise at levels greater than 73dBA produced erratic results with an IW machine and our work has confirmed this. Current CW machines are more resilient to noise and can be trained to recognise noise as an interspersion between words but they are still prone to corruption by noise which coincides with speech particularly noise of a transient nature. However, a problem which has been observed with the CW machine used in our current field trial is that noise coinciding with the end of an inputted word can cause the recogniser to 'hang' whilst it waits for the word end. This effect not only results in a mis-match but also in a slower response.

The effects of ambient noise on the machine can be reduced also by the use of noise-cancelling microphones mounted close to the operator's mouth and by noise reduction measures in the operating area. Several recent machines have the facility to measure the ambient noise level in isolation and 'subtract' this from the inputted speech. Although this technique works successfully in an environment where the noise level is constant, eg. an aircraft cockpit, it is not wholly successful in a sorting office environment where there is also superimposed impulsive noise.

The effect of ambient noise upon the operator must also be considered. Its effect is two-fold Firstly high noise levels produce fatigue and stress which will lead to reduced working efficiency. Transient noise is distracting and again will lead to lowered efficiency and errors.

Secondly and more important is that high levels of noise cause the operator to speak more loudly and thus to pronounce words differently to the stored patterns. One means of coping with the noise level is for the operator to wear ear-protectors in which case he must be provided with side-tone otherwise the ear-protectors will have exactly the same effect as the high noise level. A side effect of providing ear-protectors is that they must be provided on a personal basis for reasons of hygiene.

Dust is present in most places where mail is being handled and its effect is to produce dryness in the throat. This causes changes in the pronunciation of words and it also causes fatigue. Both effects increase errors and reduce performance.

One area which has not yet been fully addressed is the provision of feedback to the operator. In the case of keyboard operation the feed-back is two-fold. The operator has tactile feed-back from the key. There is also a lamp on the keyboard which indicates when the machine is ready to accept an input. It is extinguished when a key is pressed and is relit after a fixed delay (usually 0.75 S). There is no confirmation of the code keyed by the operator other than that gained by looking at the keyboard as the button is pressed. Keying an incorrect code results in mis-sortation of the parcel. Measurements indicate that the error rate with a standard keyboard varies between 1 and 10%. It can be argued that feedback should only be given when the machine fails to recognise the input. Alternatively that no feedback should be provided and that inputs which are not recognised should cause the parcel concerned to be directed to recirculate for reprocessing. These strategies would reduce interruption of the operator's work rhythm. Experiments reported elsewhere (3) have shown that continuous feedback tends to slow the operator since he waits for confirmation on every input but it results in vastly improved accuracy. Furthermore, visual feedback proved to be significantly faster than auditory feedback. Our work at ELPCO tends to confirm these findings particularly in respect of slowing the operator. Feedback has another function in that it gives confidence to the operator in the machine's ability to recognise his voice. His perception of the keyboard operation is much different in that it is a more mechanical action and the need for feedback is consequently diminished. Any endeavour to reduce the level of feedback must ensure that the operator's confidence in the performance of the machine is maintained.

Training the operator to use the equipment effectively and reliably is a key issue in the application of automatic speech recognition to manual handling tasks. It must be realised that the intelligence and motivation of staff employed on these tasks will vary widely as also will their age and physical abilities. The objective must be clear; it is to train the operator so that the man-machine combination is reliable and resilient over the normal periods of operation. It is essential to consider the man-machine combination since the current level of technology cannot realise a machine which is fool-proof in its operation and resilient to mis-use by the operator. It is an entirely speaker-dependent machine and the operator must be clearly aware of this. Training must address:-

  a) explanation in lay terms of the reasons for employing speech recognition; ie. reduction in physical nature of task, ie. hands-free; better productivity;

  b) How the system operates. This should indicate an explanation of how the machine operates; the controls, the display and etc. The purpose of training the machine should be explained. He should be practised in how to train (and re-train) the machine. This is an excellent opportunity to demonstrate how variations in the speaking voice can affect the machine's performance.

A similar approach can also be used in demonstrating the importance of placing the microphone correctly.

Finally the operator must be clear about what to do when the equipment appears to mis-operate.

  c) Explanation of training programme. The operator needs to have a clear idea of what is expected of him; what the training will involve and what levels of performance he must achieve. The training programme should also involve the first line supervisor. He should be made aware of the objectives of the programme and he should have a sufficient understanding of the system to enable him to sort out minor problems which the operator may encounter.

4. FIELD TRIAL AT EAST LONDON PARCEL OFFICE

An operational field-trial of speech recognition

commenced at East London Parcel Office (ELPCO) in March 1987. This employed a single Marconi Macro-speak recognition system on one parcel sorting machine. In the first phase of the field trial (March to July 1987) a total of 48000 parcels were handled. Handling rates were between 15 and 40% higher than that obtained on the other machines (using conventional keyboards). There was no evidence that the use of voice recognition increased the error rate; in fact the general opinion was that error rates were lower than with the keyboard but this has yet to be substantiated. Figs 1 and 2 show the voice recognition equipment in use at ELPCO. The microphone is supported on a light-weight headset and is a cord-less type in order not to restrict the operator's movements with trailing leads.

The trial revealed several ergonomic problems most of which have been discussed above. Handling rates in this office are generally low (in the range 400 to 700 items per hour) and the poor ergonomic design of the sorting position contributes largely to this. Although the improvements in sorting rate were of the order predicted by our experimental work the actual rates achieved are not as high as we would expect to achieve under better conditions. Since the first phase of the trial the equipment has been in continuous operational use and some of the ergonomic problems have been tackled. Also further data has been collected which is in the process of analysis at the time of preparing this paper.

The installation at ELPCO has of necessity been of an experimental nature. It is already clear that considerable modification to this installation is necessary to make a 'production' system which can be installed in any PCO. It is true to say that the proprietary systems available at the present time fall some way short of the requirements of such a production equipment particularly in respect of our type of application ie. the machine control application. Most of the proprietary systems attempt to address more than one market sector eg. control applications and data-entry (word-processing) applications for which the requirements are very different.

The key features of our application are:-

i). The system is operated by staff who are far from 'computer literate'. It is essential that the knowledge required to operate it should be minimal. The operator control panel must be simple to operate with the minimum number of controls.

ii). The means by which an operator is logged onto the system and his set of voice templates is selected must be simple and foolproof. The use of any type of floppy disk is unacceptable as they are insufficiently robust for the type of environment in which the equipment is used. The logging-on operation should be the responsibility of a supervisor via a suitable control panel.

iii). Visual feed-back is currently via a single line LED display. We are currently examining single and two-line displays to enable more information to be displayed in the re-train mode.

iv). Most currently available systems require a host-computer. In a typical PCO we would have between three and seven installations and a separate computer for each would be expensive. We are examining how a single computer can be used to serve several systems.

APPLICATION OF AUTOMATIC SPEECH RECOGNITION TO OTHER MANUAL HANDLING TASKS.

The tasks of handling heavy and/or awkward items and the entry of data to a control or computer system are mutually incompatible. The physical task produces strain and fatigue; usually requires the use of both hands and often requires the operative to be in a standing position. The data entry task involves the operation of a keyboard with one or both hands. It requires the reading of data, often with poor optical contrast and in varying positions on the item. There is a varying cognitive element dependent on the degree of translation of the data into keystroke(s). The data entry task necessarily interrupts the handling task either in the middle of the task or between items.

Voice recognition is applicable in principle to any handling operation in which the two tasks of physical handling - particularly of heavy or bulky items - and data entry to a control or computer system are combined. It clearly offers the advantage of freeing the hands for the physical task. It can however add to the cognitive task involved with the derivation and entry of the data and account must be taken of this factor.

In additon to parcel sorting a short study has been made of the application of these techniques to recording data from mail-bags relating to the receipt of these bags at a Main Letter Office.

The main conclusion of the study was that voice recognition could be used to advantage but only if the data recorded could be standardised to one or two pieces of information. With the present arrangement, which uses between one and five data words for each item, the time taken to read and enunciate is often greater than the time between successive items (approximately 5 S).

THE FUTURE

Developments in speech recognition systems is continuing apace. Systems are now available with vocabularies of 2000 and 10,000 words and with response times in the order of 0.5 seconds. However these systems carry the serious penalty of greatly increased training time. Recognition algorithms are also being improved which should result in further reductions in overall response time. This is essential if the overall recognition time (including the time for enunciation) is to be reduced. More important from a hardware point of view is the increasing availability of single board systems. When the performance of these systems approach that of the current stand-alone systems then integration of speech recognition with control systems will become possible resulting in more economical overall systems. Unfortunately the improvements in techniques and hardware have not been accompanied by a commensurate increase in application technology amongst the purveyors of these systems. It is still left to the user to solve the manifold problems involved in any application and which now tend to pale into insignificance the technological problems associated with more efficient recognition algorithms. Speech is a natural means of communication and is therefore seen as an obvious medium for data input. However it is only 'natural' between human speakers. When speech to a machine is involved we still have a long way to go to make it more 'natural'! To achieve this I would identify the following areas which require investigation -- and which we hope to address in the near future:-

i). Significant reduction in response time by the use of expert systems approach to the vocabulary search -- ie. partitioning the vocabulary on the basis of partially received speech rather than waiting for the whole utterance:

ii). Further studies into making the man-machine system more resilient to environmental noise eg. using some of the techniques mentioned above. It is considered that this work could also reduce some of the stresses such as dryness of the throat which are a side effect of speaking in a noisy environment.

iii). Several speech systems provide an indication of the accuracy of template match in the form of a 'score'. This enables a degree of post-processing to be carried out to improve the overall accuracy of the system. This is an area which we have not had time to explore.

The list is by no means exhaustive but I consider that it will go a long way towards making speech to a machine more 'natural'.

The permission of the Engineer in Chief of the British Post Office to publish this paper is gratefully acknowledged.

REFERENCES

1. Henly H.R. "The Application of Automatic Speech Recognition to Parcel Sorting". International Postal Engineering Conference. IMechE 1986.

2. Rabiner L.R. Rosenberg A.E. Wilpon J.G and Keilin W.J.

"Isolated Word Recognition for Large Vocabularies." The Bell System Technical Journal. Vol 61. No 10. December 1982 pp 2989-3005.

3. Schierick J.M. Williges B.H. and Maynard J.F "User Feedback Requirements with Automatic Speech Recognition" ERGONOMICS. 1985. Vol 28 No. 11 pp 1543-1555.

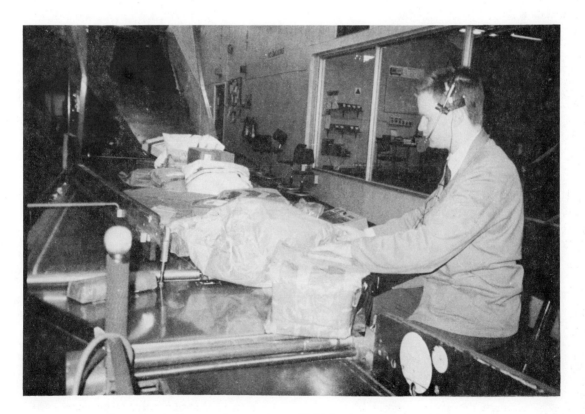

Fig 1   View of the sorting position at ELPCO using voice recognition

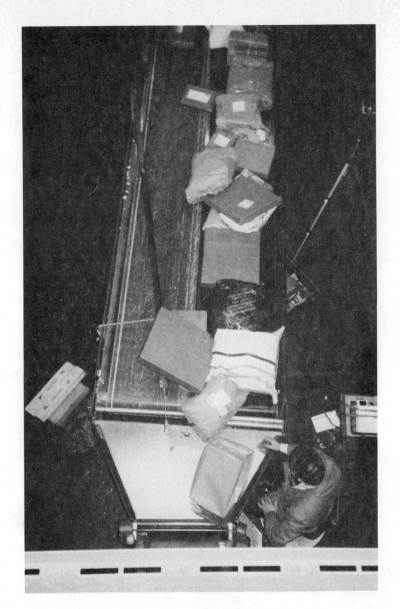

Fig 2 Plan view of the sorting position at ELPCO

C78/88

Aspects of safety in an advanced handling system

P D PARKER, FCIBSE, MILE
GEC Mechanical Handling Limited, Erith, Kent

ASPECTS OF SAFETY IN AN ADVANCED HANDLING SYSTEM

SYNOPSIS 'Safety' is a relative term and has different meanings in different situations. The use of the term 'safety' in relation to persons as distinct from damage to operating plant under abnormal conditions is discussed. Examples are given pertaining to an installation involving robots, conveyors and other handling plant where a man/machine interface is required for normal operation. The reliability of devices used to provide safety and the need to keep such devices foolproof is discussed. The paper also discusses the possibility of the creation of hazards by removing power from moving machinery in an emergency.

1 INTRODUCTION

This paper illustrates various aspects of safety in the design, construction and operation of a Flexible Manufacturing System for the production of Surface Mounted Printed Circuit Boards. Because of the use of advanced handling systems and robotics a large number of different safety features had to be considered in design.

2 DESIGN SAFETY PHILOSOPHY

In the plant described in this paper the rule was that safety should not depend on semiconductors. The robot controllers and the individual machines comprising this plant all used microprocessor based systems for sequence logic and positioning control. All the interlocking of covers, power switching and Emergency Stops were achieved by hardwired circuits operating through contactors. The microprocessor control systems were kept 'informed' of the condition of all safety circuits so that on resumption of normal conditions, the systems could restart safely.

Recently the Health and Safety Executive have published two documents concerned with the allowable use of Programmable Electronic Systems (1),(2), in safety related applications. The degree of flexibility afforded by these new guidelines is to be welcomed but for plants such as the one described in this paper the use of duplicate Programmable Electronic systems would not have been an economic choice. In any design of safety systems, reference should also be made to the Health & Safety at Work Act. (3)

3. DEFINITION OF SAFETY AND MAN/MACHINE INTERFACES

In any discussion on safety in relation to machinery there has to be a clear understanding of what the term 'safety' means. In the context of this paper 'safety' implies 'safe for humans' rather than safe in relation to plant damage. The one often implies the other but this is not always so. There are circumstances in which the operation of a safety system may cause damage to the plant in the process of avoiding injury to humans.

In the plant for the production of Surface Mounted PCBs described in this paper there are a number of Emergency Stop push buttons which open circuit breakers or contactors to cut off power. If one of these buttons is pushed during the rotary travel of any of the robots there is some danger that the energy released from the compressed air cylinders could cause overshoot of the movement and consequent damage to the workpiece (PCB). Additionally as the workpieces are held in jaws which are closed by compressed air, there is some risk that the workpiece could fall out onto the floor and be damaged.

As explained later, the robot controllers retain a memory of the positions reached in their operating sequence should an interruption occur. This feature is necessary for the manual intervention requirements of the system to enable, in normal working, the robots to resume operation from the point at which the intervention occurred. If however, the power is abruptly cut off due to some emergency whilst the robots are moving, the inertia will leave a robot in a position which no longer corresponds with the stored data. In such a circumstance, if power were to be restored without a resetting procedure, then the forces applied to move the robots into agreement with the controller's last recorded position are excessive, resulting in a likelihood of disturbance to the robot mountings. Precautions were taken to make this occurrence unlikely by ensuring that the robot controllers could distinguish between a power failure/Emergency Stop and an intentional manual instruction to interrupt the sequence.

In an automated production unit employing automatic transfer machinery linking production machines it is usual for guards and barriers to be placed so as to prevent human access to the plant whilst it is operating. There is a Code of Practice covering the construction of such barriers and the closeness of barriers to the moving mechanisms, (4). However the degree of automation of many plants is such that some human intervention is often necessary. Such human intervention can be classified under three broad headings:-

A. Necessary for production during the manufacturing cycle.
B. Necessary for setting up the production machines.
C. Necessary for plant maintenance.

Although production engineers plan manufacturing plants to operate without human intervention, in practice mechanisms fail, jam or malfunction. Unanticipated variation in dimensional tolerances of workpieces to be handled is a prime cause of malfunction of automated plant. Often the cost of designing and producing plant to standards high enough to overcome these defects would make the use of automated production facilities uneconomic. Even then, in many processes there will always be the unexpected problem requiring human intervention.

There are many processes which are, at present, uneconomic to automate completely and which require the skill of humans to allow automatic operating cycles to function.

One example that illustrates both categories A and B above is that of the automatic Silk Screen Printer which follows the pre-cleaning machine in the layout shown in Fig.1. This machine requires to be replenished with solder paste from time to time. On some machines of this type this operation can be performed automatically. However the additional cost and complexity of such a feature is often uneconomic on many designs of machine. The requirement then exists for manual intervention during a production run at periodic intervals. The timing of such intervals will vary with factors such as the size and complexity of the PCB and the ambient temperature inside the machine. This is an example of category A above.

During production the Silk Screen itself may have to be changed several times for different PCB layouts. The accuracy of registration between the Silk Screen and the PCB tracking may require to be adjusted when a change of screen is made and this process may involve manual sampling of results before the machine can be put back into the automated sequence for production. This is an example where the need arises for a man/machine interface coming within the category B above.

Another example from category B might be the machine used for automatically placing or inserting electronic components onto a PCB. Such machines have now reached the stage of development where the actual position and type of component to be placed can be selected by computer control, but such machines require to have their library of components refreshed from time to time and the composition of the component library often has to change during the working shift. In practice the complexity of machines of this type requires attention under category A in addition.

Category C above is self explanatory but nonetheless routine maintenance has to be conducted in a safe manner and often this will require machines to be operated after correction of a fault and before the machines are handed back to the production staff. There is no complete solution to the problem of a safe man/machine interface for such operations. One has to rely on the skill of the maintenance staff, backed up by training and disciplined procedures. However no maintenance manual can possibly cover all the circumstances which can arise.

4 SAFETY ASPECTS OF AN FMS FOR SURFACE MOUNTED PCBs

Figure 1 shows the layout of a plant for the production of Surface Mounted Printed Circuit Boards which embodies in its design a large number of safety criteria. The plant is divided into three 'cells' for manufacture, each containing a robot used as a transfer mechanism between plant items. In addition, other transfer machinery such as conveyors and turnover mechanisms are involved, all of which are power driven and controlled by Programmable Electronic Systems. The manufacturing process groups can be operated independently and overlap with the handling facilities which link these together. Therefore the safety arrangements require barriers which do not coincide with the manufacturing machine groups.

In this plant the workpieces (PCBs) are loaded into cassettes manually, each cassette holding up to 7 identical PCBs. The cassettes are then placed on the Cassette Input Conveyor. This is the first position where a man/machine interface occurs and safety issues arise. This machine illustrates category A above. It would have been possible at extra cost to have enclosed this conveyor, provided it with interlocking doors for access and made an interface for loading such that the conveyor was always stopped when manual loading was required. However the speed of movement of the conveyor was set at 12mm/sec and this was considered sufficiently slow as not to require complete protection from the human interface. Care was of course taken in the design to ensure that fingers could not be nipped or trapped. This conveyor places a cassette under the overhead transport system of the pre-cleaning plant which collects cassettes, one at a time as required, to pass them through the pre-clean plant. It acts as a storage device automatically feeding cassettes into position as the overhead transporter removes them. Thus it is convenient to be able to load this conveyor whilst it is moving or whilst it is stationary. The indexing action of the conveyor was controlled by photo-electric cells which detected the presence of a cassette and stopped the conveyor with the cassette in the correct position. The motor drive was fitted with a spring loaded slipping clutch and in addition an over-run limit switch was fitted. Both these devices coupled with the slow speed were considered safe enough to allow the conveyor bed to be left open and unprotected. It could therefore be maually loaded at any time without requiring any operator controls. It was of course provided with the obligatory Emergency Stop push button.

The operating sequence for safety cell No.1 which follows the pre-cleaning plant, is that the robot collects a PCB from the turnover unit and places it into the loading table of the Solder Pasting Printer. When printed with solder paste the robot collects the PCB and transfers it to the input conveyor of the Pick and Place machine.

The Solder Pasting Printer is mounted through the safety fence barrier surrounding the robot so that the silk screen can be changed from outside the safety barrier. The machine is itself interlocked such that it cannot be opened up for this purpose whilst the robotic operation is active. This arrangement allows the other items inside the cell to continue to operate whilst the silk screen change is taking place.

The safety considerations are complex because of the requirement to be able to carry out trial printing on a manual basis as well as being able to operate as part of the automated cell. Even when on manual operation the machine itself is automatic in its printing sequence and as such requires a safety cover to be closed after loading with a PCB. In the manual mode the closure of the safety cover signals the control system of the machine to commence its printing operation. However this cover is a simple unpowered manually moved item, which must not interfere with the automated operation with the robot. Thus there is an apparent conflict of requirement between manual loading and totally automated operation. In the one case the cover must be closed and interlocked when the machine has to operate, and in the other case the cover cannot be closed while the machine is operating. Figure 2 shows the robot loading a PCB into the Solder Pasting Printer.

Now consider the safety conditions applying to the robot itself. If anyone is to enter the safety cell the robot must be made inoperable. The arrangement adopted in this particular instance was to interlock the entry system of the safety cage with the robot movement sequence. The robot's programmable controller 'looks for' a signal in a number of set positions in its operating sequence which instructs its programme to jump to a parking sequence. If this signal is found the robot controller parks the robot in a safe position and remembers the original position in its overall sequence, so that when released it can resume operations from the point at which it was stopped. Merely parking the robot would not meet the safety requirements, which were that the robot must not be capable of being activated when persons were inside the safety cage. The robot parking program places the robot arm on a parking stand situated inside the safety cell in an 'out of the way' position which contains a mechanical latch and a set of electrical limit switches. As the robot arm descends onto its parking stand the latch is engaged which prevents any sideways movement of the arm, the only movement possible being vertical. Further descending movement of the robot arm actuates the limit switches which tell the controller first that the movement is complete, and secondly to cut off the power to the arm. The same limit switch system is also used to energise an unlocking latch to allow the Master Key for the cell gate to be released from its receptacle in the control panel. The action of cutting off the power to the robot also dumps to atmosphere any compressed air held in the cylinders of the robot arm. It will be seen that the robot is used to switch itself off electrically and lock itself in a safe position by its physical movement. This does not involve the semiconductor logic after the robot has parked safely, but should the electronic programmable logic fail in the process of parking or during normal operation of the robot, the system is still safe because access to the safety cage is not possible with the power turned on.

The Master Key referred to above is normally trapped in a lock on the control panel for the cell in a position marked 'run'. When entry to the cell is required the key is turned to a position marked 'run down'. The key remains trapped in this position. This position gives the signal to the robot controller that the robot is to be parked. The robot first completes any action that has been started at the time that the instruction was given and then goes into its parking routine. When the parking sequence is completed a lamp lights and only then is the unlocking latch energised allowing the key to be turned to the 'off' position, opening a contactor to remove any possibility of power being restored whilst the cell gate is opened. The same contactor removes power from both the turnover unit and the input conveyor to the Pick and Place machine, thus preventing any movement of these unguarded items when persons are inside the safety cage. It also immobilises the silk screen printer.

Thus the position reached is that the robot is mechanically latched except for vertical movement and electrically isolated so that no power to move it vertically is available. The mechanical gearing and linkages of the robot drive make it impossible to move it in a vertical direction manually.

The Master Key, now removeable from the control panel, unlocks the safety gate for the cell which can be entered. The control panel for the cell is inoperable with its master key removed but the robot's programmable controller remains operating. A secondary key previously trapped in the gate lock is released and is used in a switch mounted inside the safety cell on the Solder Pasting Printer to switch on the motive power again, but only to the Solder Pasting Printer. The actions above have informed the Solder Pasting Printer that manual operation is required and the interlocking circuits for the cover over its loading table are now connected for this purpose. These interlocking circuits also inform the system when reverting to automated operation that the cover for the Solder Pasting Printer table has not been left closed, which would prevent robotic load and unload operations.

The above description of the safety system appears cumbersome when written down because of the complex and apparently conflicting considerations of interlocking for both manual and automated operation; however in practice the system has operated well and seems 'natural' to the operating staff.

The author considers that this 'naturalness' of operation is the epitome of good design. The safety considerations are carried out as part of what appears to be logical operation and cannot easily be fooled. In software terminology the system safety is 'transparent' to the operator, who does not have to think about safety procedures or consult complex manuals to be able to work this plant.

Similar facilities were provided for safety cell No.3 where the PCBs are tested. The automatic testing facility was mounted outside the safety cell but the actual machine for connecting the PCBs to the test system was of necessity inside the safety cell and PCBs were fed to it by a robot. Because of the requirement to be able to test PCBs which are different from the production batch, there had to be a man/machine interface similar to that applying in cell No.1 for the Solder Pasting Printer.

The test equipment loading table was provided with a cover which had to be closed for manual operation of the test facility, but which had to remain open when being fed by the robot. In this case there was a further requirement that the cover should remain locked in the closed position whilst the tests were taking place. This was accomplished by incorporating an electro-mechanical safety approved latch which has a spring loaded locking action and an electrical release which held the entry cover shut until the test set had completed its work. However in this instance there also had to be a facility for setting up the loading table which could only be accomplished with the cover open. Figure 3 shows a robot removing a PCB from the loading table for the Automatic Test System.

The setting up of this loading table is carried out by trained staff. In order to move the platen, two spring loaded controls have to be turned which are positioned such that both hands of the person doing this work are engaged and cannot thus be trapped by the moving mechanism with the cover open.

Among other items for automatic handling of the PCBs, the plant contains 3 cassette turnover units of similar design to each other. These devices are used to change the orientation of the PCBs for retrieval by the robots or for process requirements. Manual access is not required to these units in operating cells 1 or 2. However, in safety cell 3 there is a manual interface for loading and unloading the turnover unit which is necessary because this item is the place at which the completed PCBs are removed from the system. This turnover unit therefore was provided with a counterbalanced cover which was interlocked such that it could not be opened for access unless the turnover unit was in the manual load/unloading position. Once opened it could not be powered as part of an automatic sequence. In this instance it was decided that closure of the cover should not be used as the signal for automatic operation to be started. The operator pushes a 'go' button after closing the cover, to inform the system that it may start this device. The signal is disabled with the cover open and the cover itself is mechanically locked by the movement of the turnover device away from its loading position. Care was taken in the design to ensure that the electrical and mechanical locking features of this cover could not be fooled from outside by an operator.

5 INTERLOCKING AND SAFETY CAGES

The robots in the system described above are protected by safety cages which have access gates. One of the features of such systems is that there are no electrical limit switches or other electrical devices attached to the cages or gates. Safety is frequently obtained by using limit switches on gates and access points of safety fences, but such switches are all too often the subject of abuse, usually for maintenance, but which get left in an abused state afterwards. Moreover switches and interlocking contacts on fences are always a weak point in electrical reliability because of the ease with which they can be abused. It only needs one contact in a chain to misbehave and the whole operation is halted. This inevitably leads to such switches being shorted out and safety arrangements negated. In the installation referred to above, none of the safety cages have any switches or other electrical devices attached to them. The locks are purely mechanical. The keys for these locks have to be inserted in the control panels which are outside the safety cages, to provide power for the moving parts inside the cages, and these keys are trapped with the power turned on. No key can be removed from the control panel until the corresponding robot or other device is stationary with its power supply isolated and in the case of the robot, mechanically latched on its parking stand as well.

The gate locks are designed such that the key used to unlock a gate remains trapped in the gate lock whilst the gate is open. The power therefore cannot be restored unless the gate is locked shut. In the case of cell No.1 in Figure 1, it will be noticed that there are two gates to that cell. Access was arranged for the second gate by providing a third key, kept normally trapped in the second gate lock. This third key could only be released when the Master Key from the control panel was used to unlock the first gate and the second key as mentioned above had already been removed and used in the second gate lock. Thus one could have access to the cell via the first gate only, or by both gates, but never by the second gate alone. The key used for the second gate was not usable in the control panel, but would however fit the receptacle on the Solder Pasting Printer to allow manual operation of that machine with both gates open. Many such arrangements are possible to fit all practical circumstances.

The access gate locks of the safety cages were attached to the metalwork by bolts having special heads, requiring a special tool for removal, making it even more difficult for anyone to negate the system.

Of course, anyone sufficiently determined could climb over the fence with the aid of a ladder, but there has to be a practical limit for each particular case as to what the objectives are. In any system of safety in plant of this nature, there has to be a balance between what is reasonable to prevent an accident and the deliberate avoidance of safety procedures and devices. The objective ought to be to make the plant safe to operate in a way which is 'transparent' to any staff involved in the operations.

The interlocking circuitry for the cover over the loading table of the Solder Pasting Printer and the limit switches mounted on the parking stand for the robot were hardwired to contactor circuits and did not depend on electronic logic for safety. Moreover the limit switches were of the 'positive break' type where the separation of the normally closed contacts do not depend entirely on the operation of springs, but are forced apart by the overtravel movement of the operating plunger or lever, (5),(6).

In safety cells Nos.1 and 3 a man/machine interface has to be used during production. The operation of the machine to which power has to be restored with the safety cage gate open, is accomplished by using a secondary key which has been released by the Master Key inserted in the gate lock. This secondary key is normally trapped in the gate lock which is designed such that the gate cannot be locked shut unless this key is present in the lock. This key has to be inserted into a locked switch on the machine concerned. This switch traps the key in the 'on' position and only restores power to the particular machine. In this way it is not possible to open a gate and leave it open with any possibility of the automated plant being started up. By retaining this secondary key, any cleaning work or other operation to be undertaken by staff who do not have special training, can be carried out in safety with the gates left open.

The Programmable Electronic Controllers used for the operation of all the plant interfacing signalling in the safety cells and the direct control of the robots, were provided with battery backed memories for retention of the overall software. These controllers were however, kept live and operating as supervisory devices when normal manual operations were taking place inside a safety cage. An exception to this condition occurs under power failure conditions or if an Emergency Stop button is pushed. Although not directly concerned with the safety aspect, it is of course important to be able to re-start the plant after such intervention. Also in safety cell No.3 the actual operation in the manual and setting up modes of the loading table for the test equipment was carried out using the facilities of the Programmable Controller for the robot, in this case using its I/O logic for interlocking the test system with the direct acting hand operated contactor controls for the motor.

6 MAINTENANCE AND POWER FAILURE

One of the complications of the system described above is that should the main power supply fail for any reason, the robotic operation and that of the plant stops. It was not considered economic to use an Uninterruptible Power Supply for this particular installation. In order to be able to gain access in such an event there has to be some override system which does not depend on the power supply for its functioning. The presence of such an additional requirement must not conflict with the normal interlocking or create a possible procedure for abuse.

Under power fail conditions it is necessary to gain access to a safety cage to recover lost or damaged work and to re-set the robot to a safe position for starting when power is restored. As the Master Key cannot be removed without power to operate the unlocking facility, another method had to be used. This was accomplished by providing a mechanical override to the unlatching feature of the Master Key receptacle on the control panel. This override was itself a key but this key was normally kept in a locked switch inside the control panel. This locked switch was in the circuit of the main contactor for the plant inside the safety cage and therefore removal of this key implied that no item could be powered. Moreover to gain access to this key the control panel had to be opened. The operating handle of the main three phase isolator for the control system and all the equipment inside the safety cage was mounted through the door of the control panel and this could not be opened with the isolator turned to the 'on' position. Thus for a manual override of the system to be accomplished, the main isolator had to be turned off, the control panel door opened, the override key removed and then used to release the Master Key from its trapped position. With the override key removed in this way, power could not be restored.

The system relies on there being only a single key set for each robotic cage and these not being interchangeable or copyable. The keys, switches and locks were chosen from well known commercial sources and are large and cumbersome items, not easily damaged.

The instruction manual makes clear that the Emergency Stop push buttons are just that and should not be used other than in an emergency. Emergency Stop push buttons were provided for each safety cell in two positions; one inside the cell and the other on the control panel, outside the cell. They are provided to meet the requirements of the Health and Safety at Work Act but should only ever be required to be used when maintenance staff are carrying out work on the equipment and then only if some emergency arises.

For major repairs or adjustment, for instance after replacing an item, it may be necessary to operate the part under powered control. All the movements can be separately operated from the control panel external to the cell. However for fine accurate adjustment it may be necessary to have a trained maintenance mechanic inside the cell. The nature of the adjustments required on this plant were such that it was not practical for the maintenance mechanic to take a portable control box into the cell and therefore a second trained mechanic stays outside the cell and operates the controls. This person can see everything inside the cell and can have one hand 'hovering' over an Emergency Stop push button should anything appear to malfunction. Normally the setting up of such adjustments is accomplished by using controls in an 'inching' mode. The The robot controller can then be made to learn the final position of any movement and these movements can be replayed and observed for checking without actually powering the robot.

7 CONCLUSION

The plant described in this paper has many other features in its design concerned with both safety and operation, (7), (8). The plant has been in operation since the end of 1985 without any mishap. The entire system can be operated by one person except for manual testing or changing a silk screen. The implications of the safety features built into its design are not noticed by the staff, who regard the actions they take for manual operation when required as 'natural'. For this plant an analysis of the requirements under the Health & Safety at Work Act was done, taking into account that no operator could be present during automatic operation. When operators are required to be inside a safety cage the robotic operation is not only electrically isolated and disabled, but mechanically prevented from being moved prior to the operator being able to enter the safety cage. Unless these conditions are satisfied the key to unlock the gate cannot be obtained and these actions do not depend on semiconductors, but are hardwired. No wiring is associated with the safety cages, or the gate locks. When resuming automatic operation the robot controller's software performs checks on each movement separately from the signals supplied to it by limit switches etc. Thus if any "temporary" bridging of any limit detector were to have been done during maintenance, or a limit detector were to fail in normal operation, the system will recognise the abnormality, freeze the action and signal a report to the controlling mini computer for the whole plant. Indicating LEDs on the control panel for each robot enable an analysis of any such faults to be deciphered. A fault tree analysis was provided in the maintenance manuals separately from the operating manuals for the system.

8 ACKNOWLEDGEMENTS

The author would like to thank GEC Mechanical Handling Ltd. and British Aerospace plc at Filton, for permission to publish the information contained in this paper.

REFERENCES

(1) Programmable Electronic Systems in Safety Related Applications. No.1. An Introductory Guide. H.M.S.O. IBSN 0 11 883913 6

(2) Programmable Electronic Systems in Safety Related Applications. No.2. General Technical Guidelines. H.M.S.O. IBSN 0 11 883906 3

(3) Health & Safety at Work Act. H.M.S.O.

(4) BS5304: 1975 Code of Practice - Safeguarding of Machinery.

(5) BS4794 Pt1:1979 and Pt2: Various dates. Specification for control switches (switching devices, including contactor relays for control and auxiliary circuits for voltages up to 1000V d.c. and 1200V a.c.).

(6) BS4794 Pt2 Section 2.2:1982 Position switches with positive opening action.

NB: BS4794 Pt2 is published as several sectional publications.

(7) Parker, P. D. Integration and Control of plant in a Flexible Manufacturing System for Surface Mounted PCBs. GEC Review 1987 Vol.3 No.1.

(8) Parker, P. D. Integration and Control of plant in a Flexible Manufacturing System for Surface Mounted PCBs, 5th International FMS Conference, Stratford-on-Avon, IFS Conferences Ltd. Nov. 1986.

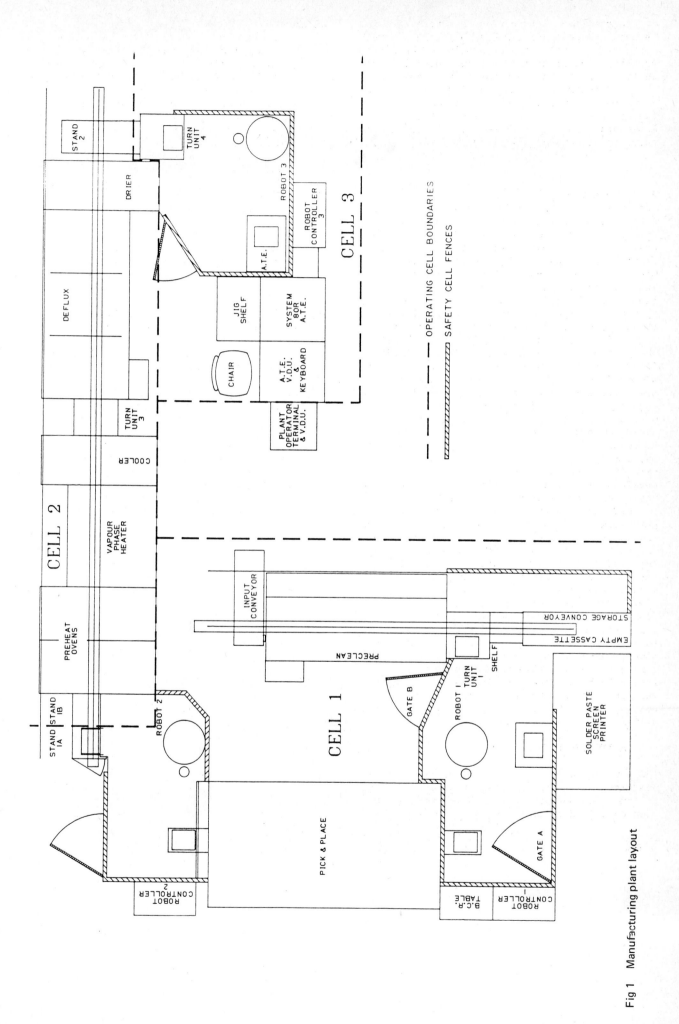

Fig 1 Manufacturing plant layout

Fig 2 Robot placing PCB into solder pasting printer

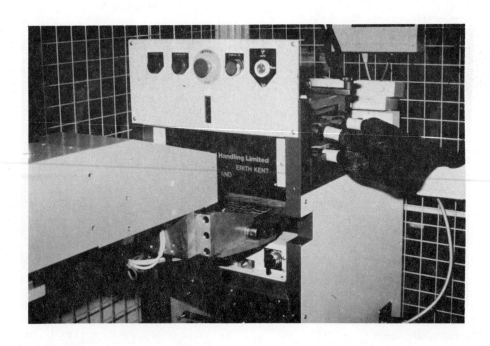

Fig 3 Robot withdrawing PCB from ATE loading table

C79/88

Elimination of parcel damage from conveyor systems

C G MacMILLAN, MSc, DIC, CEng, MIMechE
Post Office Research Centre, Dorcan, Swindon, Wiltshire

SYNOPSIS   The development techniques for gently transferring bulk mail from one conveyor to another is described in a previous paper (1). This paper describes the further development of those techniques, and a method of establishing specific designs.

1 INTRODUCTION

The Post Office has made radical changes to the design of conveyor systems used to bring parcels in bulk from the input of mechanised sorting offices to the sorting machines.

The details of the changes, and of their development, are described in a previous paper (1).

In traditional conveyor systems, at the transfer point between one conveyor and the next, the parcels are dropped from one conveyor to another as shown in fig 1. Although cushioning materials are incorporated under the conveyor belts to reduce the rise time of the impact forces, the high level of parcel interaction, both by impact and abrasion, could cause parcel wrapping deterioration and parcel damage.

The new type of conveyor system executes the transfer of parcels from one conveyor to the next without a drop, both conveyors being at the same horizontal level, as shown in fig 2.

2 CRITERIA FOR THE CHOICE OF TYPE OF LEVEL TRANSFER

Although curved belts provide the best method of producing a junction, it is not possible to use them at all junctions. There are three applications in which they cannot be used:-

(a) Where mail being fed from two directions is merged onto one take-away conveyor;

(b) Where the take-away conveyor is narrower than the conveyor feeding the junction; and

(c) Where there is an external physical constraint, such as a building stanchion, right at the internal corner of the junction.

Any of these criteria prescribe that the junction is a T-junction rather than an L-junction and the sideways feed onto a take-away conveyor has to be used.

3 BELT TRACKING

Paper (1) discusses the extensive development work which took place to alleviate the tracking problems associated with level transfer T-junctions between conveyor belts made of conventional belting materials. Because these belts can buckle, they are difficult to restrain, and susceptible to tracking off the end rollers. An alternative belt material, which is rigid across its width, is used. It is made of small glass reinforced plastic slats hinged together on metal rods so that they can articulate round the end rollers. By running this belt in a shallow trough, the belt is restrained from any lateral movement. This arrangement is very successful, and its use solves all tracking problems.

The glass reinforced plastic slats of which the conveyor is made are very durable and give very long belt life. They are however, vulnerable to damage by impact, and should not be used in any position where items are likely to be dropped onto the belt. Since the whole object of the new level transfers is to eliminate drops, this is not a problem.

The slat conveyor is run in a shallow trough to prevent its lateral movement. The trough also serves the function of preventing small items of debris, or thin items from getting under the belt. If this were to happen then the belt could be damaged as it passes over the end sprockets. It is therefore important that the sides of the trough are adjustable so that they can be moved close up to edge of the belt and fixed in this position.

4 DEBRIS IN THE SYSTEM

Some debris inevitably finds its way into mechanised sorting systems. This debris is usually associated with the mode of the arrival of mail into the office. For example, if the mail arrives in mail bags, the system is likely to contain numerous bag labels together with their associated strings and bag seals. Alternatively, if the mail arrives in containers, the occassional towing pin may be expected. To prevent such items becoming jammed in an in-running nip between the head end of one conveyor and the tail end of the next, it is important that the coefficient of friction of the belt running out of the junction is the same, or higher, than that of the conveyor feeding the junction. However, this may not be possible if the feeding conveyor is a steep rising conveyor, and has to be made of a high friction material.

To overcome this problem, a number of solutions have been developed:-

4.1 ANTI-JAM ROLLERS

An anti-jam roller device is shown in fig 3. It consists of a roller lagged with a material of higher friction than that of the infeed conveyor, and a scraper bar. The gap between the anti-jam roller and the head end of the feeding conveyor is adjustable, and allows items of debris to fall through into a debris tray. Some small items of mail may also pass through into the debris tray and the gap must be set such that the percentage of small items passing into the debris tray is not unacceptably high. The debris tray must be easily accessible for emptying. The position of the scraper bar is also adjustable, and is set so that the gap between it and the anti-jam roller is virtually zero. Minute items of debris drawn into this gap cause eventual erosion of the roller lagging, which is therefore designed to be of low cost and easily replacable.

The anti-jam roller device reverses the effect of the in-running nip. Items forced down into the gaps either pass through them, or are driven back out. It has proved to be very successful in most situations, preventing damage both to parcels and to conveyor belting. It is particularly attractive because it maintains the level transfer concept, and it automatically culls items of debris out of the system.

However, it is not so effective in systems which are fed with mail from mail bag opening positions. This mode of input causes mail bag labels, each attached to a piece of string and a metal seal, to be fed into the system. These items of debris pass over the anti-jam roller, and some of them rest on the scraper bar. They remain there until fibres of the end of the string become drawn between the anti-jam roller and the scraper bar, and cause erosion of the roller lagging.

4.2 INCLINED ANTI-JAM ROLLER

Fig 4 shows an arangement of the anti-jam roller device designed to overcome the above problems. It achieves this by having the upper surface of the scraper bar inclined so that the labels cannot rest on it, but are tipped onto the take away conveyor. It is a small departure from the level transfer concept since the take away conveyor is slightly lower than the infeed conveyor.

4.3 POP-UP ROLLERS

Fig 5 shows another design, consisting of a bed of rollers, each roller of which can easily ride up out of its seating and become released. These are called pop-up rollers. They have proved to be very effective in systems which carry a large number of very heavy items with protruding appendages which can catch in the junction between conveyors. They have the disadvantage that once a roller has been released by a trapped item, it is fed through the system. It is therefore necessary to have suitable arrangements for its retrieval and replacement.

4.4 CASCADE

A small drop, or cascade, from one conveyor to the next has the disadvantage that it is a departure from the level transfer concept, but by using small diameter conveyor end rollers, the drop can be minimal. Parcels descend in two steps, the front end toppling first, followed by the trailing end. Impact forces are thus very low. A second disadvantage is that all debris is transfered over the junction and thus passed through the system, rather than being automatically culled out.

Both these adverse effects can be reduced if it can be arranged for the takeaway conveyor to rise at its input end, as shown in fig 6. This practically eliminates the drop, and allows some debris to roll backwards off the tail end of the take-away conveyor.

5 SIDEWALL DESIGN

The design of the conveyor sidewalls on the internal corner of the junctions is very important. Using a force transducer to measure the sidethrust on the take away belt it has been found that with the correct design of sidewalls, the actual conveyor arrangement can be greatly simplified. If the sidewalls on the internal corner are not sufficiently relieved and correctly profiled, then the additional refinement of flairing out the end of the feed conveyor with a transition bed of rollers increasing in length, as shown in figure 7, is necessary to give a smooth flow of parcels through the junction. Such a refinement is mechanically complex, since the rollers are of different lengths and require to be mounted on a trapezoidal main frame, and to be individually driven by a complicated staggered drive arrangement. On a 1.5m wide conveyor system the longest rollers are 3m long, and need to be centrally supported on auxiliary idle rollers. These are potential sources of jams because they can draw in thin items of parcel debris. The correct design of side-plates renders the flairing out of the bed of rollers, with all the associated design and manfacturing complexity, unnecessary.

Design of the sidewall opposite to the incoming mail is also important. If parcels extend across the conveyor and come into contact with this sidewall, then the mail can jam in the junction. A belled sidewall which allows parcels reaching it to be easily pushed up it, prevents jams, and allows a free flow through the junction. For this to happen, the maximum angle of inclination of the lower end of the sidewall has to be 30°, and great care has to be taken to ensure that the bottom of the sidewall is lower than the upper surface of the belt. Jams can be induced by a lip as small as 2mm if this is proud of the belt. Since the conveyor material is rigid slats of finite and accurate thickness, correct interfacing of sidewalls is not difficult to achieve in manufacture.

It became apparent during the trials of different side-plate designs that there is no single design of side-plates to suit all the varieties of junction and traffic. It has been demonstrated that the following factors affect the specific design of side-plate for a particular installation:-

(a) The peak parcel throughput, in items/hour;

(b) The nature of the spread of parcels,
    ie evenly spread mail, or
        mail in slugs from tippers;

(c) The width of the conveyors; and

(d) The maximum length of parcels.

6 DESIGN GUIDE FOR LEVEL TRANSFERS

The success of level transfers is very dependant on selecting the correct arrangement for any particular application, and the appropriate detail design is very important. The Post Office Engineering Department has produced a design guide in the form of an intelligent system consisting of a floppy disc and an accompanying booklet. The designer of an installation can use the intelligent system at a personal computer, and by responding to specific questions is directed to the appropriate detail design drawings.

7 JAMS AND BLOCKAGES

The incidence of mail jamming at junctions is greatly reduced when level transfers are used rather than conventional junctions incorporating a drop. The reason for this is that parcels dropped from one conveyor to another momentarily come to rest, and then accelerate from rest in another direction. This momentary hesitation causes accumulation at the junction. It is usually necessary to have the conveyor leaving the junction running at significantly faster speed than that feeding it to prevent the accumulation causing a jam. In contrast, parcels passing through a level transfer tend to keep flowing smoothly through the junction, with no accumulation, and consequently no jams. Quite apart from the damage to parcels and to conveyor belts that they can cause, jams and blockages in conveyor systems are often a signficant disruption to the flow of parcels through the office and an additional maintenance cost since they generally have to be cleared by engineering maintenance staff. The use of level transfers thus enhances both the throughput and the efficiency of a system.

A direct comparison has been made of improvement in performance achieved by replacing a traditional junction with a level transfer in an existing system. The throughput was increased from 2,300 items per hour to 2,600 items per hour. This represents a 13% improvement in the throughput capacity of the system, and a reduction in maintenance time of over two man-hours per day.

8 RETROSPECTIVE INSTALLATION IN EXISTING OFFICES

The new standards for L-junctions and T-junctions have been installed in the latest Parcel Sorting Office at Reading and have been adopted as standard for all new mechanised sorting offices. Work has also been carried out to devise ways of installing the new designs retrospectively in existing mechanised offices. The problem in doing this is that the junctions that have to be modified are invariably in the main input system to the office, so that the disruption to the working of the office while the modifications are taking place is potentially considerable and must be minimised. Although each junction in each office must be studied separately, it has been found that in most cases it is possible to perform the modification by designing and building the extra conveyor and rollers required off-site and fitting it as a conversion kit mounted over the existing conveyor system. The actual on site installation time can thus be reduced to about 36 hours, and confined to cutting away existing sidewalls and chutes, and fitting the new conveyors. It is usually not necessary to remove, or even move, the existing conveyors or their rollers, etc. Figures 8 and 9 show typical modification kits which can be used to modify the junction represented in figure 1.

A study of all major mechanised sorting offices has taken place within the British Post Office to identify where systems can be modified to remove drops, and replace them with level transfers. As a result, a large number of modifications have been carried out. The introduction of level transfers has not only reduced parcel damage in mechanised offices, but has reduced the incidence of parcel jams and down time in the main input systems in the offices. Thus, level transfers afford a simple method of achieving an immediate reduction in operating costs, and a consequent improvement in efficiency, in mechanised parcel sorting offices

ACKNOWLEDGEMENTS

The author thanks the Engineer-in-Chief, Post Office, for permission to prepare and present this paper, and his colleagues in the Engineering Department who assisted in the actual development project.

REFERENCES

1   MacMillan,C.G. "The development of damage free level transfers" published in the proceedings of the I Mech E, Part B, 1987, volume 201.

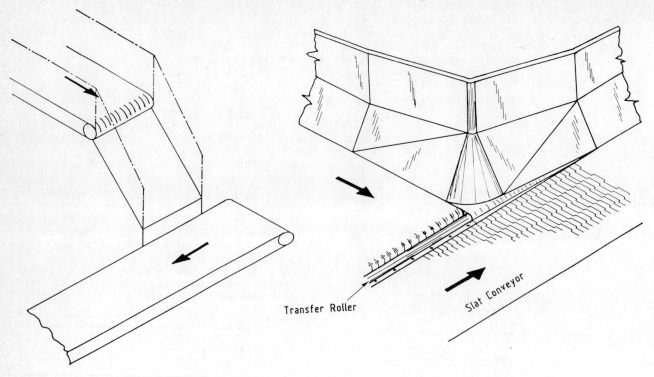

Fig 1   Traditional junction

Fig 2   Level transfer

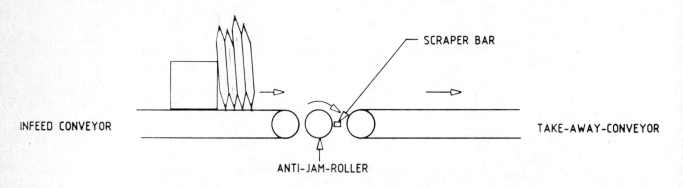

Fig 3   Anti-jam roller

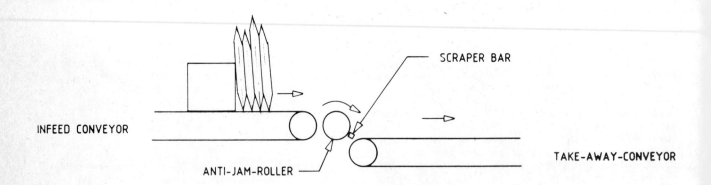

Fig 4   Inclined anti-jam roller

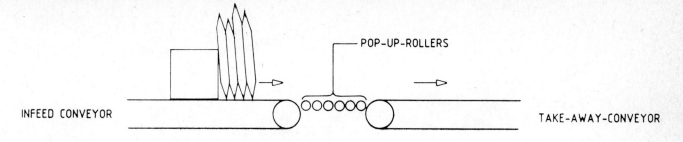

Fig 5   Pop-up rollers

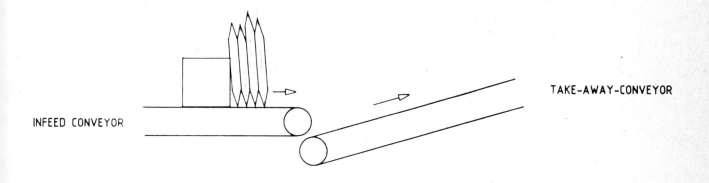

Fig 6   Cascade

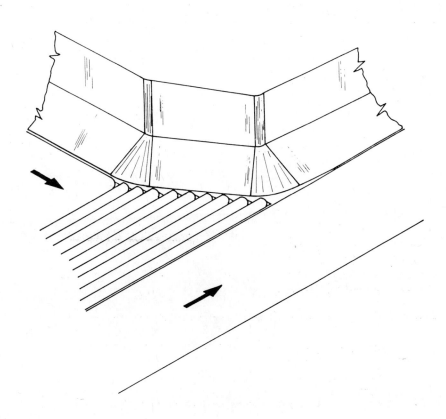

Fig 7   Flaired-out junction

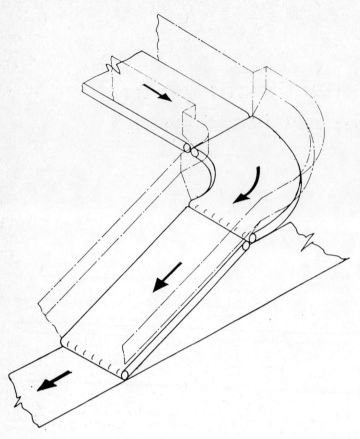

Fig 8 L-Junction conversion

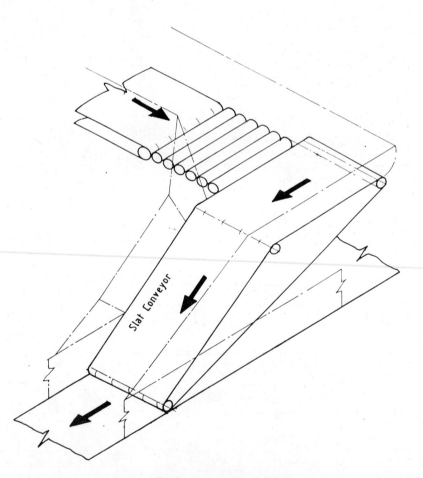

Fig 9 T-Junction conversion

C80/88

A flexible approach to manufacturing at JCB International Transmissions Limited

I A McDONALD
BT Rolatruc Limited, Slough, Berkshire

JCB Transmissions is the manufacturing base for gearboxes and axles for JCB Excavators and other companies. The plant is highly automated using a variety of proven automation technologies in manufacturing, storage and transport. This paper will describe the automated storage and retrieval systems. The factory concept, the systems integration and the investment criteria are also discussed.

JCB

JCB Excavators are one of the outstanding British success stories in post war years. From a lock up garage in 1947 to a £250 M + turnover international company in 1987.

The Companies strength lies in its ability to "do best what it knows best" and that is building excavators. But building does not mean merely assembly. It means a totally integrated approach to producing a highly marketable core product, adapted where necessary, to suit worldwide markets.

The marketing strategy extends to a competitive price whilst the design and manufacturing makes no compromises toward quality and performance for the customers. Not only earning a return on investment but also making a reasonable profit is another prime aim.

There has been no magic in this formula just hard work and a good deal of all-to-uncommon sense coupled to a foresighted investment programme and a determination to succeed. Of all the projects JCB has instigated, JCB Wrexham has been the most ambitious and most successful.

JCB TRANSMISSIONS

JCB Transmissions at Wrexham is the manufacturing organization which sells its own transmissions and axles to JCB Excavators and other companies.

From modest beginnings to the point where a major investment was required for the next step forward took only 4 years. In October 1983, JCB conducted their own in house feasibility study on how to expand the production capacity, contain or reduce direct costs and reduce inventory whilst increasing manufacturing flexibility.

The in house study was not an attempt to save on outside consultancy cost but rather an honest belief that internal resources know their own business best. Few managements have the will to release resources for this type of assessment and fewer still have the courage to face the truth of the findings and act upon them.

The systematic deliberations led to the unsurprising conclusion that the companys' strength lay in manufacture and sales of gearboxes and axles. Hardly a revelation but nonetheless a vital conclusion, confirming that the company was aiming its expansion in the right direction and allowing efforts to be concentrated on a single well defined goal.

JCB's objective as always was to maximise productivety per man by using investment in the most appropriate technology. JCB have a belief that "costs walk in on two feet" and if the number of feet can be minimised, then there are not only direct cost savings but also hidden cost savings in reduced social amenities. More importantly there is a greater chance of stock accuracy and "first time right" quality with automated plant.

In addition, high levels of the right automation mean greater managerial control over stock levels and quality. The automation provides "real time" information that would otherwise not be available. The right automation is an important aspect of JCB philosophy, all JCB investments are state of the art but are tried, tested and proven in some way. The combinations may be innovative and novel but the elements are known quantities. JCB does not believe in being a guinea pig.

JCB's future is dependant on a flexible response to customers requirements. Inherent in that response is quality, price and the right product. Total control over manufacturing as achieved by automation is the way JCB maintains these goals.

Materials were clearly identified as a major cost area. Hence the expansion was aimed at providing maximum control of materials.

The cost of inventory for example would be very high for stored finished goods. A weeks worth of stock is valued in excess of £250,000. So the U.K. manufacturing industry norm of stock turnover 6 times per year would give rise to stock of up to £2 million. By adopting flexible manufacturing techniques, JCB Transmissions customers can be supplied on a just in time basis without large stocks of finished goods.

The principle also applies further back in the production whereby parts are not put through a manufacturing process until required in the build program. In this way, stocks of raw material can be held at above just in time levels but value is not added until required.

Again, the right automation comes into play whereby processes are eliminated where possible. For example, by roller forging and forming splines, finish grinding can be eliminated, the product improved, production times cut and quality bettered, all at the cost however, of major investment.

Once all these requirements and direction of the company had been confirmed, a plan of action for the future was laid down by consultation with specialist suppliers.
BT Rolatruc were one company chosen for the integrated handling which is a key feature of the whole factory.

AUTOMATION CONCEPT

Material proved to be a high proportion of factory costs. To contain these, two main ideas were central to various schemes investigated:-

Automated Manufacturing

Automated Materials Handling

Automation was a key theme to provide the necessary throughput and flexibility with the highest possible levels of quality.

The automated handling had to provide speed, efficiency and safety whilst tracking the components in transit.

An important factor in such a total committment to automation has been the accuracy of stock control. Combined with a known build program this allows high stock accuracy to be maintained, a vital component in enabling inventories to be kept as low as possible. This aspect of automation in providing a real time inventory control is often overlooked but should be inherent in any company planning. Just in time methods or Optimised Production Technology.

Once the total concept had been established, a draft specification was requested from short listed suppliers. The brief was a tough one, in addition to meeting all the requirements of the concept, JCB Transmissions required that service be available 24 hours per day and service within four hours of call out; that the system should be able to provide a turn-key operation. In addition, the installation of the expansion project had to be integrated with the existing factory without disrupting or hindering production.

Visits to other installations were an important feature of JCB Transmissions investigations. There is little profit in simply being first at any technology and JCB had no wish to be used as an experiment by any supplier. Only tried and trusted technology was to be used albeit in a unique and innovative package. As well as visiting other automated storage and transportation systems (cranes and AGVs) such as Perkins at Peterborough and Leyland Unipart in Chorley, JCB went abroad to study machine tools and AGVs in assembly.

At Saab Scania in Sweden, JCB visited the factory where BT AGVs are used as automatic assembly vehicles. This concluded a range of visits concerning the materials handling and JCB then made their choice.

In May 1984, an order was placed with BT Rolatruc for the supply of a High Bay Rack and Automated Crane Storage system, AGV based horizontal transport systems for various materials movements, FMS servicing and automated assembly vehicles as at Saab Scania.

A 55,000 square foot building extension was planned to accommodate the expansion increasing the total factory floor area to 120,000 ft^2. Work began in March 1985.

INVESTMENT

As in all such installations planning is of crucial importance to success. Not only did JCB assess their own areas of expertise but also identified experts in other fields and accurately assessed what would be achieved with current technology. The planning phase included not just the installation of the hardware and computer control system but also the integration of the system with the day today production within the factory. The magnitude of this part of the project should not be underestimated by any company contemplating this type of installation. Effort and resources are not only required from the supplier but also in equal measure from the Client.

Return on investment is most commonly calculated for investments in manufacturing, these are well documented and easy to calculate. This is not so easy for automated materials Handling.

Often ROI is poor on face value, but a more rigorous investigation taking the materials handling system and the integration into account. Asking "what happens if we don't have an integrated system?" can highlight bottlenecks and under utilization of machinery.

The very act of seriously considering automation leads to a detailed analysis of material flow. This rigorous analysis in itself can provide useful improvements, by defining real problem areas which tend to occur as a company grows and changes.

A simple but effective part of the analysis is a flow diagram. Care must be taken to account for not just materials but also returns of rejects, the flow of packaging, return pallets etc and also waste materials generated. The JCB materials flow diagram shows which transports have been automated as well as clearly indicating how parts and product move throughout the factory. The total investment was £6.5m with a payback period of 3.2 years.

For ROI calculations many factors were taken into consideration. Precise details are not available for publication but included:-

- the total cost of manning levels, salaries, social amenities, administration, lost days etc etc.

- the cost of stock categorised into raw material semi finished goods, proprietaries and finished product.

- the cost of quality including the cost of **NOT** getting it right first time.

- the cost of internal transport.

- the cost of time for every operation.

- energy costs.

The basic philosophy is a holistic one in that total product costs are used to evaluate the whole production facility. This extends as far as the users product where for example, a slightly more expensive axle can be worthwhile in reducing build costs or warranty or allowing the product to be sold into previously inaccessible territories. The latter can occur through homologation regulations or service requirements.

MATERIAL HANDLING

There are over 400 parts handled in the system. Incoming parts go through quality control and are then conveyed through a weighing station and profile check.

At the end of the conveyor, a VDU is used to enter details of the pallet loads. From here the Warehouse Control System tracks all components through the system. Pallets are transferred from the input conveyor to the crane pallet store by an Automatic Guided Vehicle (AGV).

Three AGVs are used for distribution throughout the plant. Each AGV is fitted with a powered roller conveyor bed capable of lift/lower movements. The AGVs service almost all parts of the system including production areas in the old part of the factory. The two areas not serviced are the FMS cell and the finish goods output.

In addition to the crane store, there are a further 23 pick up and deposit points in the AGV system.

The total layout is shown in the diagram figure 1.

STORE

The crane store consists of 3 aisles of 10m high Very Narrow Aisle (VNA) racking and two automatic stacker cranes. See figure 2. Two aisles are designated for general storage of finished components such as bearings and gears. A single automatic crane and an automatic transfer car services these two aisles. Orders from the Warehouse Control System (WCS) are transmitted to the crane via a short range type of infra red communications link operating at the input/output end of the aisle.

For this type of communication, the crane is stationary and receives orders or acknowledges completion of assignment whilst loading or unloading. If parts are required from the other aisle then the control system automatically activates the transfer car which transports the crane from one aisle to the other.

The third aisle has its own dedicated crane which has a continuous communication system enabling messages to be given or received at any time not just at the aisle end. This aisle holds two types of material, general stores as in the other two aisles plus raw components. Outputs stillages or pallets are deposited on roller conveyors which are in a row on the lower part of the side of the racking.

These output conveyors feed manned clamping stations where components are clamped into machine pallet jigs. The clamping stations remove finished pieces ready for return to the crane store.

Beside the clamping stations, there are VDU terminals to allow the operators to instruct the control system that a piece is ready for transport.

FMS

The FMS cell consists to ten Scharmann FMS machines, two washing stations, a co-ordinates measuring station and a 60 location buffer store. The cell has its own AGV system, 3 AGVs each fitted with a special attachment for handling machine pallets (see Figure 3). The machine pallets are allowed to "float" a limited amount to ensure absolute positioning accuracy upon deposit. An automatic shuttle mechanism takes the raw component machine pallet into a machining centre whilst presenting the AGV with a pallet of finished components. The AGV transports the finished parts to one of four destinations, the buffer store, the washing machines, the co-ordinates measuring machine (CMM) or the crane store.

The CMM is a fully automatic cell checking parts under command from the control system so that quality is maintained continuously and automatically. Results from the CMM are fed back to the FMS control and adjust the machining centres if necessary.

The washing machines are also of course automatic and all components are washed after machining.

The buffer store is an integral part of the FMS cell and is serviced by the AGVs which deposit pallets at two levels.

The buffer store is used during the manned day shift to hold machine pallets of unmachined parts. These are then transported and machined automatically at night. Finished parts are taken back to the buffer until the clamping station are next manned.

In addition to transporting machine pallets, the AGVs also transport 80 tool magazines for each machining centre.

Within the FMS area two problems were encountered. Firstly the AGV's experienced skidding due to cutting oil and washing fluid dripping from the machine pallets onto a painted and sealed floor. Eventually this was cured by a combination of a large drip tray on the AGV and a roughened floor surface for the AGV wheel tracks. The solution highlighted an important aspect of this type of project namely that co-operation is of vital importance.

It would have been all too easy for JCB to blame the AGV's, then BT Rolatruc blame the floor specification, and both companies to blame the machine tool company or the washing machine manufacturers. The problem was solved by co-operation not confrontation.

A second difficult area was the interface between the AGV's and the FMS machine tools. Where the machine pallets are deposited a cup and cone mechanism is used for location. Obviously the machine pallet must be free to move horizontally during the lowering movements. To achieve this a floating table mechanism was supplied by Scharmann to BT Rolatruc. The tables were delivered complete but without design detail so the installation had to be engineered without design drawings. Swedish AGV's with German attachments made life difficult at times for the British Project Team.

Nowadays all such contracts have every interface as detailed as possible so that all parties know exactly what information they have to supply.

ASSEMBLY

Machined parts from the FMS area and bought in parts from the store are brought together in the assembly area. Gearbox assembly is on AGV's circulating in a rectangular loop. Five Special assembly AGV's carry the main casings clamped to a turret. The turret can be rotated to any angle and raised or lowered under the control of the assembly operators, allowing the workpiece to be positioned for the task at hand. Upon completion at that station, the AGV moves on to the next position until assembly is complete. The AGV's can cope with all the products manufactured at Wrexham.

Finished work is unclamped, palletised and transferred to the finished goods pickup point. The third type of AGV then transfers them to the despatch dock. The dock is outside under a canopy and the AGV exits and returns through an automatic door.

The AGV also returns empty pallets to the point of use and transfers other finished goods from the "old" factory to the despatch area.

CONTROL SYSTEM

The control system is based on a master/slave principle. JCB's own master computer provides the build plan details which are transmitted into the Warehouse Control System. The WCS is also linked to the FMS cell computer to maintain control of the machine centres, stock and stock movements.

The hardware used is by DEC, with a PDP 11/24 master, twin PDP 11/24s one active and one cold standby for the WCS and a PDP 11/44 for the FMS cell.

The whole network is totally integrated and provides extremely accurate real time stock control, an essential prerequisite for just in time methods.

A notable feature of the software is the housekeeping facility whereby components are automatically re-allocated during low activity times to points where they will be required. Throughout the control system parts for the FMS take priority over other functions, so for instance housekeeping would stop whilst FMS parts are issued.

For a view of the materials flow please see figure 4.

Clearly JCB has had faith in its own business assessments. There has been a certain amount of faith in the decision to automate but not as much as other companies seem to think is required. For those who say "well that's all very well but JCB have got the money and we haven't", then please consider how JCB earned the money in the first place; supply a marketable product (quality, performance, price) in a world market, support it properly and contain or reduce production costs. The costs are contained by investments such as at Wrexham. JCB Transmissions is a shining example of an effective application of advanced technology. A combination of production and materials handling technology at its best.

SUMMARY

1. JCB in-house feasibility study; "where are we going"?

2. Act on the conclusions.

3. Trusted technology from suppliers with appropriate experience.

4. Automation ROI analysis must be more rigorous and thorough than the norm, and must take all factors into consideration e.g. stock accuracy, production flexibility optimum material flow etc.*

5. Integration is a key to maximising efficiency.

* Which means high stock turnover, low cost of stock, better order response time, less work in progress.

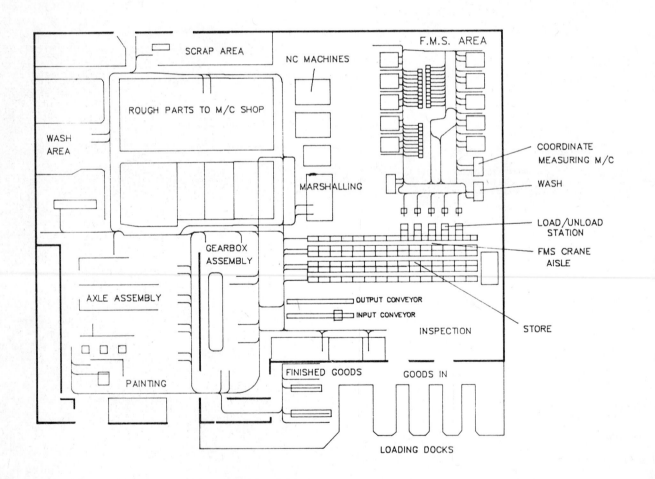

Fig 1 Factory layout showing store and AGV track layouts

Fig 2 Automatic stacker crane within the high bay racking (this photograph was taken during installation prior to commissioning the store)

Fig 3 FMS AGV showing lift mechanism, floating table and special drip tray

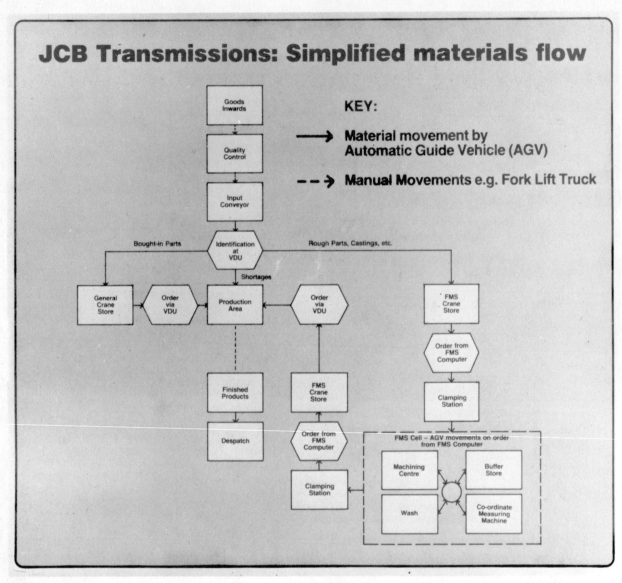

Fig 4 Materials flow diagram

C81/88

Stock handling application in a retail warehouse

A WRIGHT, CEng, MIMechE, MIMM, MInstPkg
MMM Consultancy Group Limited, Basingstoke, Hampshire

SYNOPSIS Retail Warehousing is changing, in terms of both increased specialist contractors involvement and more effective handling systems. This paper takes the reader through the evaluation, choice and introduction of two handling systems within a contractors new warehouse and discusses general descriptions and operations of each system.

1 INTRODUCTION

For many retail organisations, reserve stocks are still held in stockrooms close to the sales floor. Replenishment of the counter displays is a daily routine and when customers require an item not on show then sales staff leave the floor to find the goods from the stockroom. Over the years this situation prevailed until pressures built up for major change in the storage and distribution pattern.

Important factors included:

(a) Excessive time spent by sales staff away from floor
(b) Congestion at goods inwards dealing with intake from multiplicity of suppliers
(c) Expanding the sales floors into stockroom space.

The next stage was for retailers to move to off-site stockrooms, virtually extensions of the sales outlets. Each Branch would have a dedicated vehicle supplying new stock daily to meet immediate needs.

In a big organisation the proliferation of these outside stockrooms leads to high space and operational costs, particularly where high seasonality and the peak trading features associated with fast moving merchandise are concerned.

In consequence, as one would expect, a move from small individual stockrooms to larger, regional, warehousing has taken place, in order to obtain benefits expected from the economies of scale.

In the past few years, and because of these trends, many third party contractors have set up operations geared to meeting the physical distribution requirements of the High Street, leaving retail chains to concentrate on what they do best, namely selling.

This paper outlines a specialist contractor's warehousing facility designed to support a client retailer's requirements. In particular it describes the introduction of advanced merchandise handling equipment not previously employed in the contractor's other operations.

2 NEW WAREHOUSE AND DISTRIBUTION CENTRE

2.1 Scale of New Facility

A member company within the National Freight Consortium, commissioned a major Distribution/Warehouse facility in late 1985 dedicated to the operations of its major retailing client. The timescale covering site acquisition, build and commissioning was from October 1985 to the scheduled opening on September 1st 1986.

This new Distribution/Warehouse Centre comprises an 8000m2 consolidation depot adjoining a 16 000sq.m. warehouse, both activities providing a 24 hour operation.

The consolidation area was designed to handle some 4 000 000 packages per annum, collected from 44 manufacturers and transhipped out, either to approximately 42 local shops or to other network consolidation areas.

The warehouse has two 8000m2. bays, each bay to hold and handle either boxed, hanging or bulk merchandise in support of a number of retail outlets in London and North Home Counties.

Because of timescale and planning constraints, decisions on the building/site configuration and shell dimensions had already been made prior to detailed handling and storage evaluation.

2.2 Storage and Handling Requirements

At the planning stage of the project two issues required appraisal by the Development Team - the method of storage and handling for a significant volume of hanging merchandise - possibly up to 500 000 garments, and the corresponding facilities for boxed merchandise estimated at 200 000 cartons at peak.

A comparative cost/benefit evaluation was made for handling boxed stock between high bay storage using cab-rising order picking trucks or cranes and a multi level installation with conveyor feed/take-away facilities. Taking into account order/pick characteristics and the significant seasonality factor the multi-level approach was preferred. A similar evaluation for hanging stock also concluded multi level storage.

3 DESCRIPTION OF STOCK LAYOUT

3.1 Boxed Stock

For boxed stock a decision was made to separate bulk stock items to a future pallet handling area of the warehouse, the majority of boxed stock to be held on shelving up to 14 shelves high but serviced by both trucks (one face of the shelving) and pedestrians on Galleries (on the other face).

Initially the stock layout would be based on locating certain merchandise departments in the truck aisles with others on the Galleries, using frequency of visit and size of pick as the criteria.

The future development of the installation includes providing carton live storage between truck aisles and pedestrian aisles but this introduction would need to be tested for its impact on volume utilisation. The live storage lanes would have back-up reserve positions at ground level.

The advantages of carton live storage in this installation are that it would:

(a) concentrate order picking mainly to Gallery level
(b) reduce commitment on trucks primarily to replenishment functions
(c) improve order pick utilisation with a reduced area of pick face
(d) potentially increase cube utilisation (subject to trials).

3.2 Stock Characteristics

Boxed merchandise covers a number of different departments including shirts, mens underwear, mens shoes, ladies shoes, babywear, etc., and the range of SKUs in a typical store catalogue is significant, up to say 20 000 - 25 000.

Even at peak, stocks achieve full pallet quantities per SKU only in a small proportion of SKUs, the majority of stocks being binning quantities. This factor and the daily support to each shop in replenishing display counters down to stock quantities, i.e. singles, meant a storage system with both high accessibility and flexibility for a changing seasonal range. (These later features are also true of hanging garments.)

3.3 Storage Arrangement

The storage arrangement would therefore be based on continuous shelving installed up to full height of building and with three floor levels for access - ground plus two upper floor levels. Available height between shelves was to be such as to optimise the height utilisation. Depth of the shelving would allow 'back'-stock (i.e. one box at the pick face and one box behind in reserve).

Shelving occupies some 65 per cent of the floor area and runs transversely across the bay with perimeter aisles and a central corridor to break the shelving runs. Every other aisle between faces is a pedestrian access aisle and this occurs at all three levels. The other aisles are for narrow aisle trucks with clear truck movement to the full height of the installations.

Shelving is back to back, one face fronting the manned aisle or gallery, the other face fronting the truck aisle.

The layout comprises 18 truck aisles - aisle width 1600mm and 20 pedestrian aisles 2250mm wide.

The installation provides 14 shelves high and a total of 24 000 linear shelf metres.

The remaining area (35 per cent) is for operational tasks including receipts, checking and stock up-date (for material coming through from the Distribution Bay) stock preparation (i.e. ticketing, pricing, etc) and despatch load marshalling.

3.4 Hanging Stock

Based on the anticipated hanging stock requirement and the proposed use of space for bulky homeware merchandise in the same bay, the planned storage arrangement was to be on two upper floor levels serviced from ground level receipts.

Each floor level would have runs of horizontal storage rails either side of catwalks along which staff would perambulate.

On safety grounds, steel mesh netting infills the space between the catwalks, capable of taking a man's weight.

3.5 Bulk Homeware Stock

An increasing range of homeware merchandise has to be stored, generally bulky with heavy items typified by books, etc, and lightweight products like table lamps.

In general the storage requirement remains in shelving because of the low cube stored per SKU but with deeper and higher shelf apertures.

Storage was to be planned in two areas; the first in available ground floor space below the hanging garments 2.7m high, five shelves high; and the second in high level shelving using high rise order picking trucks 14 levels high. The provision exists for conversion to pallet storage.

The remaining open floor space would be used for operational tasks.

3.6 Handling Equipment

The handling solutions for both multi floor installations required an innovative approach, systems appropriate to a retail environment, and equipment resilient and reliable in a three shift operation.

4 HANDLING EQUIPMENT EVALUATION

4.1 Boxed Stock Handling System

The specification for a powered overhead conveyor system for handling boxed merchandise set out clear operational requirements.

Its functions were to transfer full or empty roll pallets along both a first floor and a second floor mezzanine, and from each mezzanine provide a transfer facility to and from ground floor.

The handling of existing roll pallets required a special carrier device to be part of the conveyor system. Staff should be able to easily load roll pallets onto the system and also be able to code the carriers to the ground floor destination. Alternatively, traffic moving from ground to mezzanine level should also be coded to a specific address.

The principle of the system was to be able to address and transfer a carrier to a specific location on each storage floor where roll pallets could be either removed or loaded from the carriers. (See Figure 1 for conveyor/storage relationship).

4.2 Safety

Safety was a particular requirement as the conveyor would circulate the working areas of the warehouse with staff in close proximity. Adequate emergency stop buttons were to be located around the installation and on each carrier. In addition, the leading face of each carrier was to be fitted with buffers incorporating pressure sensitive strips which cut out the supply bringing the carrier to a stop on impact.

It was particularly important that the normal activities of put-away stock and order picking should not be inhibited by the presence of the conveyor even at peak throughput.

4.3 Key Information

(a) Pallet Characteristics

A particular requirement was for the conveyor to carry safely two types of wheeled cages (roll pallets) either in the empty or loaded state.

	Cage Pallet Type 1	Cage Pallet Type 2
Height	1372mm	1759mm
Width	705mm	864mm
Depth	883mm	717mm
Weight (unloaded)	75lb	75lb
Height (floor to base)	178mm	150mm

(b) Throughput

	Current Thro'put Pallet Movements Per Shift (nom.)	Projected Thro'put Pallet Movements Per Shift (nom.)
Dayshift:		
Orders in cages)	135	270
Empty cages)	135	270
Nightshift:		
Receipts in cages)	111	222
Empty cages)	111	222

4.4 Review

A review of the conveyor proposals submitted by suppliers identified the advantages of a monorail conveyor approach over other systems namely:-

- good environmental features, low noise, oil free and visually attractive
- expansion achieved by adding more carriers
- technologically advanced over chain systems.

Detailed design was initiated particularly related to the support structure required, safety features and the actual carrier design - a variety of which were submitted in the tender stage and are illustrated in Figure 2.

4.5 Hanging Garment Handling System

A number of suppliers of specialist systems for storing and handling hanging garments were asked to submit proposals against a requirement specification which set out:-

(a) Stock capacity, to be accommodated on two upper floor levels
(b) Throughput parameters
(c) Varying length garments.

Systems available included power and free chain driven conveyors handling one metre long trolleys on which garments hung, to an approach based on linked, motor driven rods with surface spiral grooving along which the wheels of special carriers travelled.

A combination of these straight-driven-lines with unpowered bends (i.e. gravity action) produced the required movement of stock. Lateral speed of conveyors is generally 12m/min.

Each floor level was to be capable of accommodating from two to five horizontal storage bars to suit all lengths of garments but still providing easy access for staff to pick orders or put stock away. (See sketch in Figure 3).

4.6 Rapid Garment Intake

Additional features were to include provision for rapid intake of garments on receipt and for this suppliers proposed telescopic booms which would traverse progressively into a vehicle as the offloading progressed. The booms were to be fitted with conveyor rails, both to feed empty carriers to reach the offloading operation and subsequently to allow filled carriers to enter the warehouse for checking. The design of boom would allow both powered and manual operation and have ancillary lighting provision because of dark-hours offloading.

4.7 Key Information

(a) Capacity

Location	Capacity (Garments)
2nd Floor	257 000
1st Floor	202 000

Note. Storage rail to be based on 40 garments per metre.

(b) Dynamic requirement 54 000 garments per day intake.

4.8 Returned Hanging Stock

A large scale warehouse operation, handling a number of major shops has significant quantities of returns and requires either semi-automatic or automatic bagging equipment. Four major suppliers of this equipment submitted quotations and a cost/benefit study indicated that at peak - approximately 3000 garments/day returned - an automatic machine was needed.

The positioning and layout of the machine was to be such as to allow the proposed hanging garment conveyor system to supply empty trolleys to the vicinity of the machine and, once loaded, allow the trolleys to convey the bagged stock to the appropriate storage location.

5 TECHNICAL FEATURES

5.1 Hanging Garment Conveyor System

The conveyor selected is a modular system consisting of the following main elements:

(a) Driven Lines, which allow the automatic transport and distribution of trolleys in the system. The powered movement of trolleys is straight and change in direction is achieved by use of gravity rail curves, vertical lifts, declining or inclining lines and outfeeders, which 'lift' trolleys off the powered line onto gravity/manual rails. (See Figure 4 - photograph of driven line).

(b) Gravity/manual rails, on which the trolleys roll either by gravity or manual action. Change in direction is obtained by curves, switchrails or infeeders. The latter 'puts' the trolleys from the gravity/manual rail onto the powered line system.

(c) Miscellaneous components which include:
Trolley stops (manual or automatic release) which stop and hold trolleys in predetermined positions.

Stepfeeders which space trolleys for optimized utilization of the system capacity.

Brakes which automatically control and decrease the speed of trolleys on gravity rails to a reasonable level.

Collision guards which control the 'trolley traffic' where trolleys may collide due to T-junctions.

(d) Complete control system - located in a central panel with all necessary equipment for control of motors and supply of compressed air to the system. All cables and pipes for compressed air are led from the control panel to the outlet points in a closed cable tray integrated in the system.

(e) Incorporated in the control system is an 'emergency stop switch system' operated by pushbuttons positioned regularly on central positions in the system. A push on any button will cause an immediate 'power off' and all powered or driven functions will stop immediately. This emergency stop will not, however, disconnect the supplies to the micro and feeler switches.

(f) The trolleys, which are driven on powered lines or roll on gravity/manual rails and are directed/controlled by the components described earlier. The trolleys transport the goods around in the system. Each trolley is equipped with a coding device, which, depending on the pattern coded, will ensure that the trolley is 'lifted off' the driven line system to a gravity/manual rail at a dedicated location. (See photograph of storage lane addresses, Figure 5).

5.2 Capacities

The system capacity is governed by the driven/powered components. The following is a list of maximum capacity and recommended loading of these units.

a) Driven/powered lines:
Maximum linear speed of trolleys
12 metres/minute
Minimum space between trolleys
1 x maximum trolley dia.
Recommended space between trolleys
2 x maximum trolley dia.

b) Trolleys:
Maximum loading:
20 kg (approx. 40 lbs) per trolley.

Trolley requirement : 1900 total.

c) Declining or inclining ramplines:
Maximum linear speed of trolleys
24 metres/minute.

The following principal specifications apply:
Motors: 250W, 3 phase, 380/415 V
Control system: 24 V, a.c.
Pneumatic system: Minimum pressure 8 bar.

5.3 Detailed Description of Components

(a) Driven/powered lines:
Drive Unit : The rotating tube, internally equipped with longitudinal stiffeners for strength and fixing of special bearing-units is supported on an extruded aluminium profile, which also serves as a carrier for miscellaneous equipment.

A 250 W electric motor drives the universal tube through the worm-gear and four V-belts.

Bearing unit: On each end of the shaft is fixed a special rubber clutch for fixing into the universal tube.

(b) Gravity/manual rails:

Straight rails and curves : Patented rolled steel profile with electro-galvanised surface. Curves in angles from 15 degrees to 180 degrees. Radius of curves from 300mm to 1500mm.

Manual switchrails: Straight or 90 degrees. Mechanical unit in gravity/manual rail systems, which, manually operated, will lead the trolley to alter direction either sideways or straight.

Manual stop: Installed on the gravity/manual rail. A hook stops the trolley rolling on the rail. The trolley stop is released by a manual pull on the steelwire.

(c) Miscellaneous components:
Stepfeeder: Installed on gravity/manual rails with a V-shaped arm positioned approximately 50 mm over the rail. The stepfeeder is a pneumatically operated component.

The V-shaped arm, catches the trolley topwheel, whereby the trolley is stopped and fixed. The wheel is pneumatically released and pushed forward by an electrical pulse from the integrated control system. The pulse is normally generated by a time-relay in the stepfeeder-cabinet but may also be generated by a switch.

Pneumatic stop: Installed in connection with the stepfeeder, the stop acts as inlet control to the stepfeeder. The arm stops the trolley and releases pneumatically when an externally generated signal is received.

Pneumatic brake: Installed on gravity/manual rails. The cylinder presses the brake-rail against the fixed rail on the opposite side, whereby a pressure against the nylon-roller on the trolley is obtained, and the trolley speed decreased. The pressure for the cylinder is controlled by switches ahead and after the brake.

(d) Outfeeders and Infeeders
Pneumatic outfeeder with Codereader Device: The outfeeder allows the transfer of trolleys from driven/powered lines to gravity/manual rails.

The switches on the codereader 'reads' the pattern of code knobs on the trolley, and

if the same, will signal the electronics on the outfeeder to activate the solenoid valve. Pressurised air is then sent to the cylinder which pulls down the 'pick-up' arm. The topwheel on the trolley will now roll onto the 'pick-up' arm.

When the trolley passes a switch the electronics will signal the solenoid valve to release the pressure to the cylinder and the 'pick-up' arm will return to its normal position. The trolley will, rolling on the topwheel, turn 90 degrees away from the driven line and enter the gravity/manual rail which is directly connected to the 'fixed' end of the 'pick-up' arm. When the trolley passes a switch beside or over the gravity/manual rail, the whole system is re-set and a new trolley can enter.

Manual infeeder: The infeeder allows the transfer of trolleys from gravity/manual rail to driven/powered lines.

The trolley will, manually pushed, enter the infeeder rolling on the topwheel on the gravity/manual rail, which by the shaft, is connected to the infeeder.

When the trolley wheel reaches the 'in-feed' arm, this will, due to the load from the trolley, tip down and the angle-wheels are now at the correct level for catching the driven line. When the trolley, transported by the driven line, leaves the infeeder, the 'in-feed' arm will return to normal position with the help of a counter-weight arm.

(e) Control System
Control Cabinet; Contains the control system with all necessary relays, contactors, switches, etc. The cabinet is divided in two sections, one high-voltage section three x 220 V and a low-voltage section 24 V, AC.

The high-voltage section contains the main switches, automatic fuses and terminals for external cables.

Low-voltage section contains transformer, relays for control and motors, time-relays and terminals for internal cables.

Start/stop-panel: Control panel is located in a central position in the system. This is a remote control of the normal start/stop switch in the control cabinet.

Terminal boxes: Fixed in central places in the system and containing terminals for connection of multi-core cables, routed in the plastic cabletrays, from the control cabinet to single-core cables to specific components. The terminal box may also contain time-relays for collision guards, etc.

Emergency-stop push-button: Easily visible red push-buttons fixed in central places in the system. A push on any of the buttons will immediately switch off all power, and the button has to be manually released to enable a restart of the system.

Collision guard: This is a device, which by switches activated by the trolleys and time relays, controls the 'trolley-traffic' and avoid collisions in T-junctions or where trolleys otherwise may collide on driven lines.

(f) Trolleys
The four-wheel trolley is a transport device with a chassis of cast aluminium. The trolley consists of the following main parts:

Top part, with the topwheel and four angled wheels

The trolley body, with coding knobs, sliding on a rail. On the body is fixed a code scale visualising the actual code of the trolley.

Product carrier, which is specially designed for the loads, fastened to the trolley body by a shaft protected by a nylon sleeve/roller.

5.4 Boxed Stock Conveyor System

The conveyor system consists of the following main elements:-

(a) Extruded Aluminium Track - either straight or with 180 degrees and 90 degrees bends, and with four live busbars for 42v operation, allows the automatic transport of trolleys in the system. The track incorporates right and left hand divergence sliding switches for transfer horizontal between tracks.
(b) Drop/Lift sections with 8m of lift height, for the vertical transfer of trolleys from ground to upper levels.

(c) Power Trolleys, capable of traversing along the track at either of two speeds. Each unit comprises one loading and one idler trolley, a load bar (to assemble the load carrier), a control panel and a peg setter.

(d) Carriers, the frame which carries the roll pallets around the system. (See Figure 6 for artists impression of carriers in system).

(e) Lift Tables, motorised, cam operated, platforms 1000 x 2000mm dimensions, which can be raised to allow roll pallets to be withdrawn from the carriers. The location is termed 'Bus Stop' and trolleys will wait at a bus stop until an operator activates the table. (See operating description).

(f) Control System located in a central panel with all necessary equipment for control of motors etc. Control concept based on electro-mechanical relays with programmable logic controllers operating on 110volts, single phase, 50Hz.

A mimic board identifies and tracks the location of each trolley in the system.

5.5 Capacities

The capacity of the system depends on the speeds of operating, the number of load/unload stations, and the quantity of trolleys in the system.

(a) Trolley speeds :
    Two speeds   24 m/min
                            48 m/min
(b) Drop/Lift Sections :
    Two speeds   5.5 m/min
                            33.0 m/min
(c) Lift Tables :
    Raise or lower time   3.4 sec
    Dwell   20.0 sec

(Dwell is the time the trolley will wait at the bus stop before moving on - it can be varied to suit the operation).

(d) Maximum Weight of Load : 250 kg.

5.6 Detailed Description of Components

(a) Extended Track : The track is made up of straight and curved sections to produce a circuit and is attached to the supporting steelwork by 'C' brackets. For control purposes and to avoid trolley collision the track is electrically divided up into blocks, each block capable of being progressively monitored by trolleys through one of the busbars so as to assess whether a trolley is on the preceeding block.

(b) Trolley and Carriers : Trolleys are fitted, as standard, with an electro-mechanical front bumper stopping system to isolate the drive motor and apply the brake. The carrier also has front safety bumpers with a cut-out device to stop the trolley on impact, plus an emergency stop button. Because the track circuit occupies operating space at mezzanine levels with staff perambulating the binning, the trolleys slow to the lower speed at designated parts of the circuit for staff to cross the trolley path.
The 'bus stop' facility also incorporates an automatic control of forward travel to provide in-line queuing clearance between trolleys.

(c) Drop Tables ('Bus Stops') : Each bus stop position has an operator console to operate the raise/lower facility and cancel the trolley destination code. Also fitted are indication lamps and emergency stop buttons.

(d) Drop/Lift Sections : The drop sections are fitted with a load bar (length of track) on which the trolley rests and is effectively lifted/lowered.
The lift activity is at maximum speed until it trips the limit switch on the mast which activates the drop section switch to slow speed. The section continues until a second set of limit switches stops the motor. At this point the moving track lines up with the fixed track and the trolley with carrier will leave the drop section.

6 SYSTEM OPERATIONS

6.1 Hanging Garment Handling Equipment

All hanging merchandise will be received via purpose designed telescopic booms. (See Photograph, Figure 7).

Vehicles with densely loaded stock will reverse to one of two allocated bays. Stock is already assembled on sets - approximately 200mm long - and the vehicle driver transfers stock on sets onto the empty garment carriers already hanging on the telescopic boom.

Warehouse staff count the intake of garments by sets, agree with the consignment documentation, and code the trolley for one of a number of checking (booking-in) lanes. Lanes are pre-allocated to department or store.

Warehouse staff ensure that adequate empty trolleys, nested in three's, are available on the empty trolley lanes and as required feed these through to the end of the boom. As offloading proceeds so the driver inches the boom further into the vehicle.

Discrepancies between the set count offloaded and documentation requires immediate action. If the set quantity is divided in the vehicle load then all this stock is offloaded to a query lane for later reconciliation.

The system enables trolleys to travel down consecutive booking-in lanes if the lane coded is full. As a safeguard against mis-coding, a colour tag is attached to each trolley corresponding to the appropriate stock type.

6.2 Checking

Prior to physically placing garments to stock, intake is checked and booked onto computer stock records. Each garment type is checked against supplier order and confirmation of receipt actioned.

Movement of stock off trolley should not normally be necessary but the conveyor layout does allow for empty trolley parking to facilitate some physical transfer between trolleys if queries arise.

Checking procedures are the same for store returns.

6.3 Transfer to Storage

Stock is held at first and second floor levels and will be held in designated zones by department i.e. dresses, men's suits etc.

The receipts checking procedure will initiate the system to provide a storage address from the stock file and the appropriate trolleys with a specific garment type will be coded with the advised address.

(Once trolleys are coded and physically transferred onto the driver (powered) line, the transfer to the appropriate floor is then automatic).

6.4 Coding

Each trolley has four moveable indices for coding, from the lowest code 1.2.3.4 to the highest 8.9.10.11.

(a) The first index code identifies the floor i.e. 1 - ground, 2 - first floor and so on.
(b) The remaining codes identify the address lane on a floor.
(c) If the last code is set on 12, this signifies an empty trolley, irrespective of other codes.

The conveyor 'enters' and leaves alternate storage lanes from and to the driver line - so an individual address covers two storage lanes. The entry of trolleys to the lane is governed by a three-way switch which can be set either open, closed or in a position to only receive empty trolleys coded 12.

6.5 Put-Away

Stock will transfer to the designated address by way of driven lines and via inclined ramps. At the correct address the outfeeder will activate diverting the trolley and load off the driven line onto the manual rail and into the storage aisle.

As a precaution against mis-coding, the colour tag will accompany the stock appropriate to the type of merchandise.

Staff resources at this floor level will be allocated relative to the intake of stock and will transfer garments from the trolleys to the static storage rails in numerical or other sequence. Empty trolleys will be recoded 12 and transferred back onto the driver line or specifically coded to an empty trolley accumulation lane at the receipt area. Mis-coded trolleys will be re-coded to their correct address.

6.6 Order Selection

Computer raised shop orders will be printed in the warehouse and will be issued for picking against an agreed schedule of stock deliveries. Picking resources will be assessed and orders distributed to allocated staff.

To allow for conveyor journey time and stock preparation prior to despatch, both the timing of issuing of orders and order picking performance are important. Orders issued to staff have pre-allocated despatch lane codes already printed - i.e. designated store lane.

Once orders have been picked onto the trolleys, a security tag is attached denoting the store. The trolley is coded as advised for the lane and pushed onto the driven line for automatic transfer to the despatch marshalling area.

6.7 Despatch

A despatch team handles the influx of orders on trolleys which now accumulate on the series of despatch lanes.

In accordance with the vehicle delivery schedule, stock is transferred from the trolleys to mobile garment racks, wrapped, labelled by store, and load assembled. These racks are loaded to the vehicles and are directly employed on the sales floor replenishing the display rails.

6.8 Boxed Stock Handling Equipment

The overhead monorail system comprises two loops, one operating at each of the two upper levels along the length of the central walkway of the boxed storage system. The monorail loops are linked to the ground floor by two drop sections (elevator/lowerator). (See Figure 8 for conveyor routes).

Following receipt of boxed stock from the adjoining distribution sortation bay and booking onto the computer stock files, stock in roll pallets is taken to the ground floor load station for transfer to storage accommodation. At this stage it is only necessary for the load station operator to know which of the floors the stock is held on.

6.9 Transfer to Floor

The monorail trolley carrier enters the load station (Drop section 3) in slow speed, the operator then lowers the drop section by pushbutton, opens the security latches on the carrier and if necessary removes any empty roll pallet. He loads the carrier with a loaded pallet, closes the latches to secure the pallet and raises the drop section by pushbutton.

The operator selects destination by pressing either of the two illuminated floor buttons followed by the release button. The carrier leaves the drop section in slow speed and proceeds to the elevating drop section (No.2).

When drop section 2 is in the bottom position the loaded carrier enters in slow speed and stops in position on the load bar. The drop section now rises in fast speed to within 400mm of the floor level destination and moves into position in slow speed.

6.10 Traversing the Floors

When in position, the carrier leaves the load bar in low speed and after clearing the bend proceeds at fast speed to the lane switch SW1 a. The switch automatically moves to its straight through position and the carrier proceeds

The carrier trolley will alternate between fast and slow speed when approaching and leaving gangways or switches.

6.11 'Bus Stops' (Lift Tables)

The carrier will enter 'bus stop' (LT1) in slow speed and will stop above the lift table and wait for 20 seconds.

If the carrier load of stock is not required by the local staff then the carrier will automatically leave the 'bus stop' in slow speed and will proceed to the next bus stop and wait there 20 seconds and so on.

If the stock is required (because it is normally stored in this part of the warehouse) the operator will initially press the 'hold' button before the 20 seconds have elapsed, followed by the 'raise' button to lift the table to its raised position which then supports the carrier base.

The operator then opens the latches on the carrier and removes the loaded roll pallet. The table is then lowered to its bottom position and the operator wheels the roll pallet to the storage aisle where goods will be 'put-away' to shelving. (The operator reverses the lift table procedures when loading an empty roll pallet back to a carrier).

The carrier is then released by activating the Ground floor illuminated pushbutton followed by the release button. This sequence operates the peg setting cam and the carrier leaves the lift table area. Once past the zone clearance area the peg setting cam is retracted in preparation for the next carrier approaching the lift table.

6.12 Orders

Store orders will be picked directly into one of two types of roll pallets, depending on the merchandise, by perambulating the shelving. When a roll pallet is filled or an order complete the roll pallet will be parked at the nearest 'bus stop'. The operator continues picking but when a carrier stops at the bus stop will hold the carrier then assemble the roll pallet on the lift table.

The lift table will be raised and the pallet securely located within the carrier. The procedure will then be repeated so as to transfer the pallet load to ground floor.

When the carrier approaches further 'bus stops' en route to the drop section the carrier passes a peg reader. If the peg has been set by a previous operator the hold button will be inhibited and the 20 second dwell will not apply, the carrier continuing through the lift table position.

(If a 20 second dwell occurs and an operator fails to press the hold button the carrier will recirculate the floor before returning to this 'bus stop' position again).

6.13 Despatch

At the ground floor station, roll pallets with orders will be removed from the carriers and marshalled by vehicle run.

The roll pallets used for order picking and transfer on the system are used within the stores for replenishment of the display counters.

7 SUMMARY

The two handling systems described represent economic and flexible approaches to merchandise handling in a retail warehouse.

They share the virtue of simplicity but have potential to meet changes in throughput or product characteristics. In particular the monorail system, normally used in a production environment, was the first installation in a retail warehouse.

There were minor difficulties during commissioning primarily related to operator error and detailed monitoring of operations was necessary. Weekly records of downtime, spares usage and operating hours were introduced to cover the full year (2000 hour) warranty period. This is essential in a non-production environment where levels of maintenance support is minimal.

Both systems offered enhancements:- a live storage system for hanging merchandise to increase capacity and extension of the monorail system to other boxed stock areas.

The clients' view was that they would not hesitate to install the same equipment in any future second warehouse.

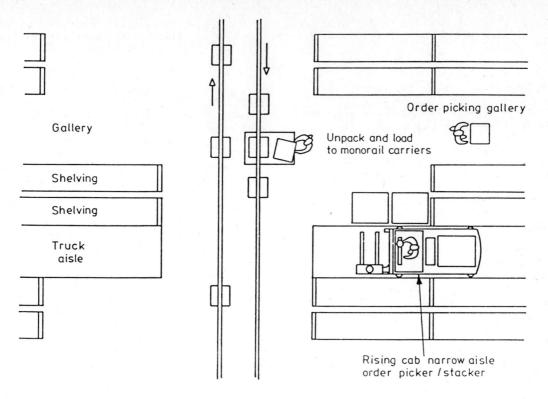

Fig 1　Conveyor/storage relationship

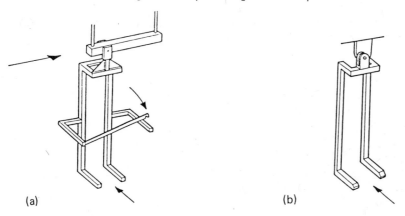

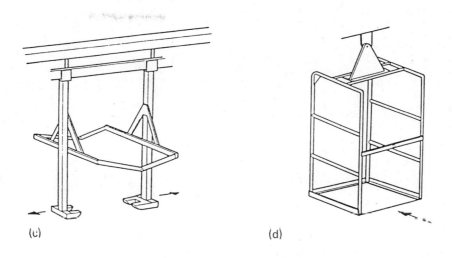

Fig 2　Alternative trolley carrier designs

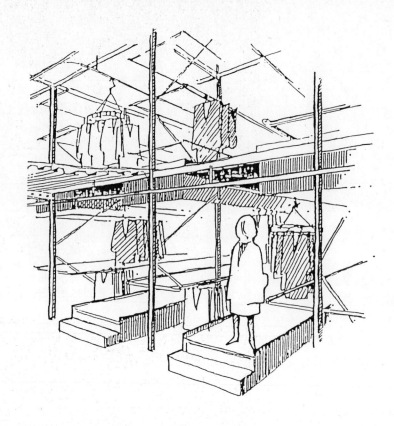

Fig 3 Hanging garment system showing catwalks

Fig 6 View of monorail system at first floor

Fig 4 Driven line hanging garment conveyor

Fig 5 Address notation

Fig 7 Intake telescopic boom

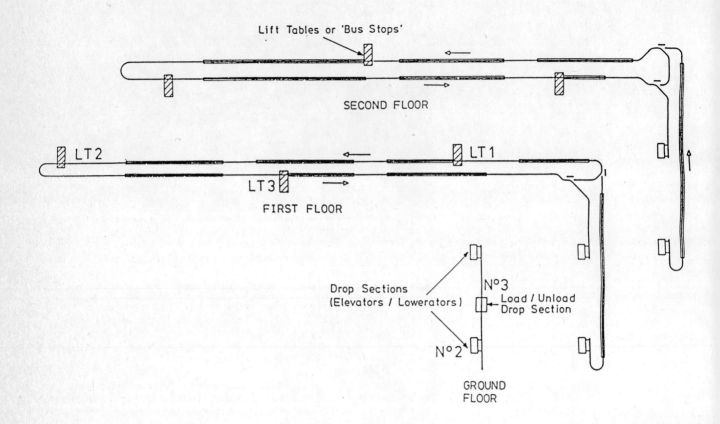

Fig 8 Floor layouts of monorail conveyor system

C82/88

Warehouse control systems

P D SPERRING, MBCS
SattControl UK Limited, Aldershot, Hampshire

SYNOPSIS The functions of a Warehouse Control System are dictated by user requirements and mechanical design. However, all Warehouse Control Systems have to control and co-ordinate the operation of various items of automatic materials handling equipment. In addition, they usually operate in conjunction with the user mainframe computer, but should also be capable of operating in a stand-alone mode. A working installation is described in an illustration.

WAREHOUSE CONTROL SYSTEMS

When talking about Warehouse Control Systems, the term 'Warehouse' is taken in its broadest context to include both distribution and production warehouses.

The requirements of each are basically the same, but the operation of a production warehouse is usually more time critical than that of a distribution warehouse. This is because the material stored is required for some manufacturing process and if the material in question is late being delivered, then invariably this has a potentially serious effect on production through-put. However, in a distribution warehouse, operations are not usually so time dependant although distribution managers may disagree.

Whichever type of warehouse is considered, the functions of the control system are basically the same. These can be all, or a selection of, the following:

- to keep an accurate and up to date record of those goods stored in the warehouse.
- to control the input to, and output from, the warehouse.
- to select and give orders to automatic cranes.
- to select and give orders to Automatic Guided Vehicles (AGVs).
- to interface with local controller for conveyors, palletisers and other equipment.
- to control order picking operations (on man rider cranes, at a forward picking area or on manual trucks via a radio link).
- to display management information on VDU terminals and printers in the form of lists and reports.
- to take orders from, and report to, the company mainframe computer.

The Warehouse Control System is usually based on a minicomputer with a second as a standby. It communicates with decentralised microcomputers in the automatic cranes, AGVs, and manual trucks and with the local controllers for conveyors, profile gauges etc.

This design is adopted as some functions are best suited for holding centrally (e.g. the database, overall equipment supervision, mainframe communications). For reliability, control of the individual items of material handling equipment is then placed either on the equipment or as close to it as possible.

EQUIPMENT

The mechanical handling equipment in an automated warehouse has already been mentioned but only briefly. Some clarification is now required.

A crane runs on rails with or without a transfer car. It can be fully automatic, picking up pallets from pallet stands or conveyors for input to the store and depositing on the same devices for output. If order picking is required then the cranes can be semi-automatic, where the crane operation is automatic and the order picking is performed manually with pick instructions received via an onboard computer terminal.

An Automatic Guided Vehicle (AGV) is used for transporting pallets of goods to and from the crane pick up/deposit positions from the loading dock. In a production warehouse, AGVs can also be used for transport between the store and production, and between production areas. The AGVs can be fitted with forks for picking up loads from the floor or low level racking, or can carry loads on the top.

THE CONTROL SYSTEM

For ease of development, implementation and after sales support, a warehouse control system should be modular in form with processing distributed to as low a level as possible. For example, the crane movements (horizontal, vertical and fork control) should not be carried out by the central computer but at a local level by a microcomputer on board the crane. Similar functions should be carried out by microcomputers on board the automatic guided vehicles.

Each of the modules within the control system should contain as much standard software as possible. This will enable the development time to be kept as short as possible and gives the benefit of at least part of the control system being based on software that is already in existence and working.

Between the modules, the responsibilities are shared as follows:

- Crane Control

  The central computer communicates with the onboard microcomputers and provides transport orders which can cover input, output and aisle transfer movements. The microcomputer then controls all the crane movements and, on completion, sends an acknowledgement back to the central computer.

- AGV Control.

  This comprises task control (transport orders), traffic control (the route between the pick-up and deposit positions) and truck control. Task control is carried out by the central computer, traffic control by decentralised microcomputers and truck control by onboard microcomputers.

- Conveyor Control

  This is accomplished by use of Programmable Logic Controllers (PLCs) which control the more basic part of the operation (for example start, stop, interlocks, monitoring, etc) under instructions from the central computer.

- Inventory Control

  The central computer maintains records of all goods in the warehouse, whether they are on shelves, on pallets in the racking, being transported or handled in any other way.

  Inventory Control maintains data about the goods, both static (part number, name, unit weight, etc) and variable (quantity in stock, reserved quantity, oldest pallet, youngest pallet, etc). Data is also held for each pallet (part number, quantity, storage date, etc) and each location (pallet number, whether location is blocked or not).

- Order Handling

  Output requirements, whether for full pallets or for order picking, are handled by this module. Requirements can be input via a VDU terminal or transmitted from a mainframe computer. The module then selects the most suitable pallet for output according to the retrieval rules (normally FIFO - First In, First Out).

- Communications

  In addition to communicating with various microcomputers, the warehouse control system may also communicate with a mainframe computer. This module handles the transaction and maintains log files on disc for each type.

DEVELOPMENT OF THE WAREHOUSE CONTROL SYSTEM

Once a company has decided to install an automated warehouse, it is essential that a clearly defined design document is prepared. This is the functional specification and is usually prepared by the supplier in conjunction with the end user and must be agreed by both parties before system development commences. It is important that the functional specification is a detailed document both for the end user and the supplier. It is the document on which the eventual system will be based and will be invaluable during the commissioning and acceptance testing stages.

The functional specification is also the document on which the next development phase is based. This is the preparation of a system specification, which is the detailed design work for the software. It includes module and program specifications, flowcharts and interface requirements. Coding of programs unique to the installation, and any required modifications to standard software, takes place in parallel with the system specification development.

Phase three, once system development has been completed, is the testing of the whole system on the supplier's premises. This involves running the system under artificial conditions where crane movement, AGV movements, etc are emulated by the system and obviously follows on from the supplier's own in-house tests.

We now come to the installation phases, where mechanical handling equipment (cranes, racking, conveyors, etc) are delivered to the end user's site and are erected/installed. In parallel with the above, the computer system and peripheral devices are also installed on site.

Overlapping the installation phase is system commissioning where, for the first time, the system is connected to and tested with the mechanical equipment. Providing that, as previously mentioned, the warehouse control system is modular, then, as installation of each mechanical handling module is completed, system commissioning of that module can commence. (For example, once the racking and cranes have been installed and tested, commissioning can commence even though, perhaps, an AGV system interfacing with the cranes has not yet been installed).

Once each module has been individually commissioned, the linking of those modules, and commissioning thereof are performed.

The penultimate phase is the supplier's own total system test where the entire system is checked for performance against the functional specification (and any subsequent variations).

Once the supplier has completed his system tests, the end user (in conjunction with the supplier) commence acceptance testing. This, obviously is an extremely important phase as, until it has been satisfactorily completed, the end user will not have his warehouse and the supplier will not have his final payment. It is, therefore, of utmost importance to both parties that the method of acceptance testing should have been clearly defined (and agreed) at an early stage.

It can be seen from the above that the various stages of commissioning can take a long time, and it is not unreasonable for six months to be allowed for these stages.

Training of the warehouse staff who are going to operate the system is obviously important and this should take place parallel to the final stages of commissioning so that they are able to assist with the acceptance tests.

POST IMPLEMENTATION

Following acceptance of the facility by the end user, the supplier should provide a period of support on hardware (both mechanical and computer) and software. Mechanical hardware support usually takes the form of a warranty from the supplier with, possibly, some of the end user's own maintenance staff trained in first-line maintenance by the supplier. Computer hardware support is normally provided by the computer manufacturers.

While hardware support is comparatively straight forward (either a component works or it doesn't), software support is required, to be able to track down those obscure problems that inevitably arise in any computer system. The quickest method of accomplishing this, assuming that the software supplier does not provide a permanent site presence, is via a telephone modem link. One telephone call then gives the oportunity for the supplier to log on to the system, via a VDU terminal, to establish the cause of the problem, and provides access to on-site experts, if required.

PERKINS ENGINES AS AN EXAMPLE

At Peterborough, Perkins Engines has a components handling facility which includes:

- an automated high bay store for stillages (five cranes, 11200 locations)

- an automated store for pallets (three cranes, 3600 locations)

- a semi-automated crane store for order picking (six cranes, 6400 stillage locations and 6400 tote locations).

- conveyors

- Automatic Guided Vehicles (20 off)

- a forward picking area (Break Bulk - 7 picking stations).

The Warehouse Control System for the installation has a PDP 11/84 as the central computer with a PDP 11/44 as a cold standby. There are three communication lines for interfacing with the company mainframe computer and a telephone modem link for software support. Connected to the central computer are various devices including laser scanners, VDU terminals, printers, industrial terminals, distributed microcomputers for the cranes and AGVs, and a conveyor controller.

Incoming goods, in metal stillages or on wooden pallets, are put onto a conveyor system in Goods Inwards. This conveyor system includes two main infeed conveyors each of which has an identification station.

As a stillage or pallet travels along the infeed conveyor, it passes a laser scanner where the permanent barcoded stillage/pallet label is read. This enables the warehouse control system to determine the type and size of stillage or pallet being used.

The stillage or pallet then travels to the profile gauge (controlled by the local conveyor controller) where the size of the stillage or pallet is checked to ensure that it will fit into the relevant store. If the size check fails, the conveyor controller notifies the warehouse control system which prints a reject note. This is inserted in the stillage or pallet which is then taken to the reject conveyor.

Stillages and pallets which pass the profile check are identified into the mainframe computer and details (including part number, quantity and destination) are transmitted to the warehouse control system which prints a movement ticket for insertion in the stillage or pallet.

Normally stillages and pallets will contain only one part, however those destined for the Pick Store can contain more than one (either packed or in totes). Each of the parts will be identified to the mainframe computer and details of each are transmitted to the warehouse control system for individual movement tickets to be printed. The details of items to be stored in the Pick Store will also include aisle number, storage zone and an indicator to show if the part is stored in a stillage or a tote.

When the identification operation has been completed, the operator will release the stillage or pallet which will then be routed to a store, inspection, reject or despatch. If the stillage or pallet is routed to a store or despatch, a transport order will be issued to the AGV system otherwise the destination is given to the local conveyor controller. In all cases, the conveyor controller will instruct a shuttle car to pick up the stillage or pallet and deposit it onto its destined conveyor spur.

In the event of the mainframe computer not being available, input can still continue with identification via VDU terminals connected to the warehouse control system. The information is saved by the warehouse control system and transmitted to the mainframe computer when communications have been restored.

A similar function exists for parts

arriving from the production area, but with less automation. Here, the stillage or pallet barcode label is read by the operator using a barcode wand, not by a laser scanner. All other functions remain the same with the exception that all stillages and pallets are transported to their destination by AGV.

The AGV route takes the AGVs past two laser scanners. The first is passed by all AGVs carrying stillages and pallets from Goods Inwards and those carrying stillages from the High Bay Store. When the AGV drives past the scanner, the unique barcode label on the stillage or pallet is read to confirm its identity.

The second scanner is passed by all AGVs carrying stillages and pallets from the Break Bulk stations, input from production, the pallet store and the pick store. Again, the function is to confirm the identity of the stillage or pallet.

Stillages and pallets destined for one of the automatic stores, that have not originated from Goods Inwards or have been input from production, are first taken by AGV to a profile gauge.

There are two profile gauges, one for pallets, the other for stillages. At both, the AGV drives in, deposits its load, waits for the check to be carried out, then picks up the load again and takes it to its destination.

The profile gauge checks the dimensions of the stillage or pallet and load. This is to ensure that the automatic cranes do not attempt to put an oversize load into the racking.

If the profile check fails, then the stillage or pallet is taken to the reject conveyor.

THE AUTOMATIC STORES

As previously mentioned, there are two automatic stores within the system.

The High Bay Store

The High Bay Store is 50 metres long, 18 metres high and is equipped with five automatic cranes (one per aisle). It is used for the storage of metal stillages which come in three sizes - small, medium and large. All sizes can be stored in aisles 1,2,3 and one side of aisle four. In these aisles, two of the smaller sizes can be stored in a location instead of the larger size. The other half of aisle four and the whole of aisle five are dedicated to the small size only. This arrangement permits the optimum utilisation of available space.

Each aisle has two stillage transfer cars, one for input and one for output. These were supplied because of the differing stillage sizes and transfer stillages between the AGV and the crane on input, and the crane and AGV on output. Each stillage transfer car can hold two stillages one at the crane pickup and deposit position (P&D) and one in the midway buffer position.

Stillages are sent to the High Bay Store, on AGVs, from Goods Inwards, input from production and the Break Bulk Area. All stillages going to the High Bay are checked by Profile Gauge to ensure that the stillage and contents are not oversize.

The stillages are transferred to the crane by a stillage transfer car and the crane places them in the location decided by the warehouse control system according to the following:

- if the stillage is a large one, then it is placed in the nearest available location to the crane P&D.

- if the stillage is a half size one, then the warehouse control system remembers the latest location where a half size stillage was stored for this part number.

- if the latest location was a front location, then the stillage will be placed in the nearest free rear location to the crane P&D. If no rear location is available, the stillage will be placed in the first free front location.

- if the latest location was a rear location, and if the receipt date of the new stillage is within a certain number of days of the receipt date of the stillage in the rear position, then the new stillage will be put in the front position. If the time difference between the two receipt dates is too great then the new stillage is placed in the first available rear position and the old stillage is taken from its location and put in front of the new one.

When a stillage has been stored, an acknowledgement is sent to the mainframe computer.

The stillage remains in The High Bay Store until an output order is issued. These output orders are either initiated by transactions from VDU terminals or by the mainframe computer. Output stillages may be routed to Despatch Break Bulk, Inspection or the reject conveyor. Stillages are automatically routed from the High Bay Store to the Pick Store for replenishment purposes.

The Pallet Store

The Pallet Store is 75 metres long, 8 metres high and is equipped with three automatic cranes (one per aisle). It is used for the storage of goods on wooden pallets.

Pallets are sent to the Pallet Store, on AGVs, from Goods Inwards, Break Bulk and input from production. All pallets are checked by a profile gauge to ensure that the pallet and contents are within the maximum dimensions permitted.

The pallet is placed on a parallel bar P&D position by the AGV and the crane places it in a location as close as possible to the P&D position. When the pallet has been stored, an acknowledgement message is sent to the mainframe computer. The pallet remains in the Pallet Store until an output order is issued. The output order is initiated via a VDU terminal or by the mainframe computer and is for delivery to Despatch, Break Bulk or Inspection.

The Semi-Automatic Pick Store

This store consists of 6 aisles each 75 metres long and 8 metres high. Each aisle contains a pick crane which has a man onboard who performs the picking operations and controls all fork movements. The crane can either be manually driven by the operator or automatically driven by the onboard microcomputer. The pick cranes are in constant communication with the warehouse control system.

Each crane is equipped with:

- a display unit showing the current position

- a display unit showing the destination location

- two lamps to indicate which side of the aisle to access

- a small terminal with keyboard and display for operator acknowledgements

- a barcode wand to confirm location and for reading barcode labels

- a printer, for the printing of movement tickets.

The store holds both stillages and totes. Each location is marked with a barcoded location number.

Each aisle is equipped with stillage transfer cars as per the High Bay Store.

Goods for input to the Pick Store are transported, in stillages, by AGV. Each stillage may contain one part or several parts in separate totes. When the stillage has been picked up by the stillage transfer car, the crane operator receives an order to collect it. If the stillage contains one part,

- if that part is stored in stillages, the operator will be given a stillage location

- if that part is stored in totes, the operator will be given the first replenishement location. When replenishment is complete, the stillage will be stored in a stillage location.

If the input stillage contains more than one part, these will be in totes and the operator will be given locations where those totes will be stored. When the operator arrives at the given locations, he will read the location barcode label using the onboard barcode wand and follow instructions given to him via the terminal.

When performing an output operation, the crane operator will first pick up an empty stillage, then follow instructions given via the terminal, picking items from the given locations and putting them into the stillage. Once he has completed the operation, the stillage will either be taken to the output stillage transfer car or will be stored within the racking awaiting call-off.

Forward Picking Area (Break Bulk)

When the quantity required is less than that stored in a stillage or pallet, the stillage or pallet is taken to one of the Break Bulk stations to be split. For convenience, either the picked from or the picked to stillage or pallet can be issued and the other returned to store.

Order Handling

The majority of orders, both for the Pick Store and the automatic stores, are transmitted to the warehouse control system by the mainframe computer. Facilities exist within the warehouse control system for orders to be input via a VDU terminal.

Pick Store orders are issued with a pick start date which may be amended. The warehouse control system will queue the orders for picking as soon as the start date is reached or passed. Picked orders are stored in the Pick Store until called off, except for some special types of order which are output as soon as they are picked.

Picked orders held in the Pick Store and orders for output from the two automatic stores must all be released, via VDU terminals, before the material is issued.

Shortages are raised by the warehouse control system and automatically fulfilled as material becomes available.

Reliability and Availability of Warehouse Control Systems

The reliability of a warehouse control system depends on using as many standard items as possible. This applies equally to hardware and software.

Computer hardware reliability has improved significantly over the past few years and the design of a warehouse control system should ensure that as many devices as possible are from one supplier. Interfaces between the computer and peripherals should also be as standard as possible.

Software reliability depends on the amount of standard software modules used by the supplier. An experienced supplier of warehouse control systems will have built up a library of standard software, based on many similar systems. There will always be software developed to suit a specific installation, but the use of as much standard software as possible increases the reliability of the total system.

Availability depends, to a large extent, upon the design of the warehouse control system according to the requirements of the end user. Availability can be maximised by the use of dual disc updating, installation of a standby computer and modem links to the software supplier.

The standby computer can be such that it is continuously checking the 'live' computer and automatically takes over should the other fail. This, however, is expensive and most systems have a second computer ready to take over once the data files and peripheral devices have been manually switched over. This means that recovery is accomplished within minutes rather than seconds but the operation of most warehouses is such that this is acceptable.

C83/88

Software implementation procedures for advanced handling systems computer control systems

A St JOHNSTON, BSc, CEng, FIEE, FIERE, FBCS, FIRSE
Vaughan Systems and Programming Limited, Ware, Hertfordshire

Software is an important element in complex Advanced Mechanical Handling Systems. Software is usually purpose written for a project and thus needs disciplines for both its specification and implementation. A procedure is described that has been used on a number of real time computer control projects.

1 INTRODUCTION

Advanced Handling Systems (AHS) usually have computer control in order to implement complex sequences in real time. Although software packages are becoming increasingly available for standard functions such as stock control, the design of AHS schemes is usually unique to each application, and thus requires application software to be specified, designed, written and tested for the job. The management of AHS projects is usually under the control of personnel from a different discipline, or even non-professionals. As a result it is often the software which causes problems, whether of time scale, cost or incorrect functions. Procedures have evolved to minimise these risks. There are many different approaches; this paper describes one that has stood the test of time in the AHS and other on-line real time control environments.

2 COMPUTERS IN AHS

Computers play a part at all levels of control in Advanced Handling Systems (AHS). Programmed Logic Controllers (PLCs), dedicated stacker crane and Automatic Guided Vehicle (AGV) controllers are provided by equipment suppliers as an integral part of their supply, and the software for these items is a package. This is the Level 3 of control, Level 1 is the corporate Mainframe. The Level 2 control is the 'mini on site' that integrates the Level 3 controls; interfaces to Level 1; services plant operating positions such as VDUs; and provides local management information. Due to the uniqueness of each installation, a software package is usually not a possible solution. The software for Level 2 is thus tailor made to fit the project requirements. This means that the particular suite of software is novel, and there is always a risk in implementing novel software, as is widely acknowledged.

2.1 Software Implementation Procedures

Procedures will not guarantee success in implementation because of the many openings for human frailty in systems of this complexity and magnitude. However, a discipline is essential. Figure 1 shows a typical overall sequence of events in an AHS project.

Due to the special nature of software implemented functions, the Level 2 computer control system is often not considered in detail at the requirement stage as are the mechanical elements of the system. However, as the Level 2 computer integrates the different mechanical subsystems of AGVs, conveyors, stacker cranes and so on into a homogeneous system, it is an area which has to be well defined, and which must also be implemented in a systematic manner if a project is to be a success.

As has been emphasised in the paper 'The Specification of Advanced Handling Systems' by R. H. Hollier and G. Shimmings, it is vital to specify the functions of all parts of an AHS in a requirement specification. This must cover the software of the control system in requirement terms, but not in computer function terms. After the contract is placed, the first action should be to generate a detailed definition of Control System functions. This document is prepared by the supplier in close collaboration with the customer and his consultant, if applicable. It is identified as a Software Functional Specification (SFS), because it should specify the software in functional terms rather than in computer programming terms.

It is important that the Software Functional Specification is complete and unambiguous as far as is possible. This should be the only document that the software team needs, apart from standard computer literature. If the Specification is not clear, is wrong or impractical, it must be discussed and updated. Unrecorded agreement between user staff and programmers is an easy route to trouble because the SFS is the document to which the Level 2 system's function should be implented, commissioned and accepted. If no such documentation exists, management of a project is almost impossible, especially if subcontractors are used.

3 SOFTWARE FUNCTIONAL SPECIFICATION

Over many years a standard format of SFS has proved capable of fulfilling this role. It is now stylised into a number of chapters.

The Introduction should describe the overall

system, including a drawing of the mechanical layout in schematic form, in order to give an understanding of the overall AHS function in the plant, but not to be a definition as regards the functions. The schematic diagram should identify in particular the pick up and put down positions of product transfers from one system to another.

Chapter 2 gives the computer hardware configuration showing VDUs, printers etc; the connections to Level 3 subsystems such as PLCs and Controllers; and any link to a Level 1 mainframe. This is not a purchase Specification for equipment, but a statement for the guidance of the software team. Hardware specification is covered later in the paper.

Chapter 3 is Data Definitions. This chapter specifies the form and range of all values, parameters, descriptions, status and so on, such as 'The Product Code is a 6 figure numeric of the form 123456'. Computers are internally purely numeric in function and it is therefore necessary to ensure, firstly that all values, status and so on are correctly coded, and also that the customer agrees with this. Often different departments have different ideas on the codings of the same thing.

The next chapter identifies the Files of the Level 2 system. Preparation of an SFS is a design activity. The file structure in an AHS usually includes a large stock file, production schedules, delivery schedules and the like. The structure of these with their associated indexes will often determine whether the response time of a system is adequate under real time conditions.

3.1 Functional Chapters in the Software Functional Specification

There then follows a series of functional chapters based on the principle that a computer process is primarily activated as a result of some input. It thus should be specified forward from sources of input, of which time is one, and should not be written 'backwards' from a required output. Those chapters may thus be titled Automatic System Functions; VDU Operator Procedures; or Actions on Incoming Messages. The chapters are tailored to fit the particular system, but should follow a pattern. Each chapter should specify all the actions that follow a particular functional area, whether as the passage of time, such as 'at end of shift...', or as a result of an operator keying a command or an input value; or as a result of an unsolicited message from another system. Polling is an automatic system function because it is carried out routinely.

The final chapters define aspects of the system such as message protocols to, say, AGV controllers, cranes or mainframe.

3.2 Figures and Appendices of the Software Functional Specification

The text should be well supported with appendices and figures which detail, for example, VDU screen formats and printed layouts, preferably on squared paper, rather than as a computer printout or a drawing, because descriptive comments can only be clearly differentiated from screen data or printed data on a drawing, rather than a printout.

In an ideal world, the Software Functional Specification should be prepared before the software is quoted for. This is usually not possible, but it should be done without fail as the first software activity after the contract is placed. Programmers should not start work until the document is finished, or largely finished, and is agreed with all parties. Preparation of an SFS is part of the software supplier's design job, but needs participation, information and meetings with the customer, and also other suppliers, particularly in a multi-supplier or multi-subcontractor situation.

3.3 Software Design and Modularisation

When an SFS is agreed, the detailed software design and modularisation process begins, which is the first activity of the software project leader. There are a number of modularisation techniques well covered in the literature, such as Jackson's method and Structured programming. One that has been used for over 25 years is called MACE (MAster Control Executive). This provides a real time structure as well as a modularisation framework. It is also designed for safe shared file usage by many processes, i.e. multi-task. AHS software is not multi-user where elaborate arrangements are made to deny shared usage.

4 HARDWARE SPECIFICATION

The computer hardware is often determined primarily by the desire to standardise on a particular manufacturer's range. However, the choice of a suitable machine in the range available requires knowledge of the sizing and access time of memory, the number and type of connections to other equipments, and the speed of the processor. Operating system and language also need an experienced assessment rather than a choice of convenience.

A dual hot standby computer system is usually cost effective. In this arrangement the files are duplicated in a second independent processor. This avoids maintenance cover being necessary out of normal working hours and avoids the possibility of losing large volumes of data. Automatic changeover is not usually necessary as a break of minutes is acceptable in the AHS environment. Other benefits from having a dual system are given below.

Hardware purchase can usually be initiated soon after contract placement, although this would be safer with an agreed SFS available. However, if ordered at the start of the Contract, the computer hardware often becomes available at much the same time as the software design is complete, and thus the software team can start on the implementation proper using the real equipment.

5 SOFTWARE IMPLEMENTATION AND PROGRESSING

Software progressing is well known to be a difficult area to manage. A long established practice is to avoid asking for, or believing in, percentage completions of software modules. Thus when the software design is complete and the modules identified, a tick off list should be prepared. This can be of the form shown in Figure 2, where the boxes for each module are S Started, D Designed, W Written, T Tested. The software modules are grouped such that firstly the group implements a particular function which can be demonstrated for program progress measurement, and even payment. Secondly, each new group adds to the previous group so that testing is a progressive identity, not carried out on isolated modules. If this is done, by the time later groups are being tested the earlier ones are much used and any faults shown up can be cured progressively during the development phase. Each module has boxes which are ticked only when completed. The Started box is to show that the module is being thought about, but only that. Tested means tested as a module within the group. Each group can reasonably be allocated a percentage of the project as a whole. Strict discipline is required for this approach to work properly, but the outcome is a reassurance to management that progress has been made and checked. Documentation should, of course, be part of the design, not an afterthought. The design of the program should be done at the flowchart or equivalent level, and not by sitting at a VDU. A flowchart should give clear cross references to the subsequent coding, whether high level languages are used or not. Assembler inserts are sometimes needed in AHS for special critical areas or protocols that are not well served by commercial operating system functions.

5.1 Simulation of Plant Controllers

During the software implementation, it is highly desirable to simulate the Level 3 subsystems. AGV and crane controllers can usually be simulated by VDUs. However, PLCs are a different matter, firstly because they use strange protocols, and secondly because their functions are to input large numbers of individual states, and output a number of Yes/No controls, these essentially being of a binary nature.

To minimise the risks and speed software testing, a real PLC should preferably be connected to the Level 2 computer. This can be without full digital I/O arrangements, but must have an input and display device for setting and monitoring PLC registers.

6 SCENARIO TESTING

Functions of an AHS as specified by an SFS are usually complex with many things happening at once, and with clever strategies to implement. 100% testing is impossible and this fact must be accepted. However, fault finding and software debugging on site are difficult for many reasons. The equipment is now spread over a wide area; the environment is not conducive to this form of activity; and the specialist back up is often a long way away. Thus, it is prudent to carry out as comprehensive testing as possible before delivery.

A test is only valid if it is repeatable. Thus a scripted scenario test procedure should be applied. There will be a number of individual scenario tests required, based on major functions. The scenarios must usually first 'set the scene' with a given file and plant status, then script a realistic sequence of events at VDUs, AGVs, etc. each with a predicted outcome according to the Software Functional Specification.

Testing one's own work is never a good procedure. Thus it is desirable, if not essential, that the user's staff prepare the scenario tests. This is a serious and non-trivial activity. Often only by preparing scenarios does the user learn how his system will work in practice. Also, importantly, whether his interpretation of the Software Functional Specification is different from that of the programming team; or only occasionally, that the system just 'will not work' unless changes are made. The presence of a clear SFS is very helpful in resolving situations at this stage without acrimony.

6.1 Joint Testing at the Factory

A good technique is to have a period of weeks or months, depending on the project size, when joint testing is carried out by both software team and user in co-operation. This spreads problems, allows thought and time for corrections to be made, and is an invaluable familiarisation period for user staff. It does not alter the need for operator training on site which, however, is best done by user staff who have participated in joint testing and are familiar with the Software Functional Specification. VDU training of operators can be catered for by providing a training package version of programs that have passed joint testing and allow VDU screen procedures to be exercised in an off-line manner.

This approach is only possible when a fully dual computer installation allows one computer system to be installed on site while the second remains on the supplier's premises.

Retention of one computer at the works also has the advantage that fault finding and program development can be carried out without interfering with plant operations. The specialist support is also more readily available at the works.

Automatic changeover is usually an overkill in an AHS system which can afford to be down for minutes while a changeover is made, but not for days while maintenance and file recovery processes are carried out in a single computer system.

The individual scenarios should test exhaustively areas of function such as product input, delivery, management functions. These will also be used to form the basis of the 'everybody play' scenario. This scenario is to test interaction and time sharing. The main activities of the individual scenarios are carried out effectively in parallel, randomly interspersed with a number of operators at different positions. If the individual scenarios have been well tested, and have no faults, this system test should not result in failures of function, but may show up failures of time sharing, and also whether the overall function will have an acceptable response in real time when loaded.

A formal factory acceptance test should be carried out, if all is well, at the end of the joint period. This should be almost a formality. It is carried out by the customer alone, but he should have largely seen it all before during the joint period. It is important, however, both because of what can be shown up, and also as a final confidence builder.

7 COMMISSIONING ON SITE

With a dual computer system, some site testing should already have taken place as regards interconnection with the mechanical equipment controls. However, no one should underestimate the time taken for on-site commissioning. The mechanical equipment has almost certainly not been continuously run with full loads, the software of the Level 2 system will have faults that are only shown up under real operating conditions. Operators will make new mistakes that the system cannot cope with. This period may therefore take some time before the system as a whole is sufficiently robust, both mechanical equipment and computer system, to allow serious continuous operational running.

A further dilemma arises at this stage. The AHS cannot really be tested under load unless that load is provided by genuine throughput, i.e. production. However, production must not be held up if failure occurs. These are directly at odds, or at best over hopeful, and a careful plan of action that caters for supplier and customer needs must be made without emotion at this point. Goodwill is essential.

8 HANDOVER AND WARRANTY

Often handover is difficult to identify as a date because the system runs for longer and longer until it is usable, albeit with some known faults that are not disruptive. A formal handover date, however, must be agreed as forming a part of the warranty period, which is usually 12 months from the handover date.

It goes without saying that software documentation should exist at this stage, with perhaps only the last changes left to be incorporated.

During the warranty period, it may be agreed that further issues of the software will be made, either to clear up residual faults, or to incorporate the inevitable changes that are only shown up during operation. At this stage only those modules should be changed which are absolutely necessary; rewriting should be avoided unless absolutely essential for speed up purposes. A new issue of software must be regarded as a risk and will require an overt trial period. The SFS and main software documentation must be kept up to date.

Before the end of the warranty period, arrangements should be made for the supplier to be held on a software field support contract, unless the customer is capable and willing to take over that job. There are established procedures for software field support in the AHS field.

9 CONCLUSION

Complex real time software is of its nature difficult to specify unambiguously, difficult to test satisfactorily, and impossible to prove unequivocally correct. However, procedures have evolved over many years to minimise the risk, and it is prudent to employ them. There are a number of methods. This paper describes one which has been successful in the AHS environment.

REFERENCES

(1) HOLLIER, R. H., SHIMMINGS, G. The Specification of Advanced Handling Systems. IMechE Conference 1988

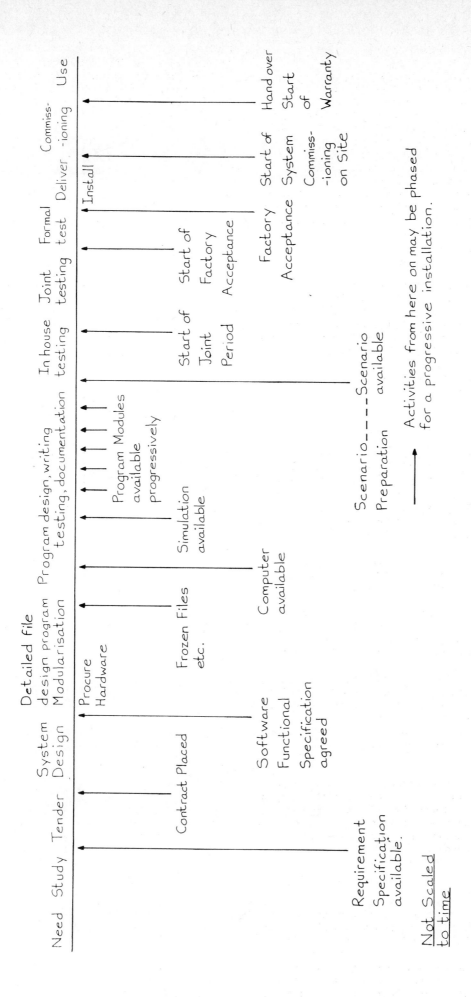

Fig 1 Project sequence

	S	D	W	T

External Load Control (5%)

E1	External Load Schedule	E1SCHEDULE	✓	✓	✓	
E2	External Load Control	E1CONTROL	✓	✓		

Despatch Control (10%)

D1	Despatch Schedule	D1SCHEDULE	✓	✓	✓	✓
D2	Despatch Drop Detail	D1DROP	✓	✓	✓	✓
D3	Despatch Load Detail	D1LOAD	✓	✓	✓	✓
D4	Despatch Confirmation	D1CONFIRM	✓	✓	✓	✓

General Purpose Area Control (3%)

S2	Stock Control	S2CONTROL	✓			
G1	GPA Control	G1CONTROL				

Highbay Control (3%)

S1	Highbay Control	S1CONTROL	✓	✓	✓	✓

Trailers and Carriers (4%)

T1	Trailer Control	T1TRAILER	✓	✓		
C1	Carrier Pallet Control	T1CARRIER	✓	✓	✓	

Equipment Status and Control (4%)

Q1	Control Screen	Q1CONTROL				
	Crane statistics report	RSTATISTICS				
	Other routines to be defined					

Reports (6%)

R1	Report Control	R1CONTROL	✓	✓	✓	
	General reports, Printer 2	RGENERAL	✓			
	Shift reports, Printer 5	RSHIFT				
	Consignment notes, Printer 4	RCONSIGN	✓	✓	✓	✓
	Despatch Office, Printer 3	RDESPATCH	✓			

Fig 2 Software module tick-off list

C84/88

Integrated control of automated handling systems

M E O'SHEA
Dexion Limited, Gainsborough, Lincolnshire

Integration of the handling system with the remainder of the business, and all subsequent levels of integration, must start with the identification of the objectives and the definition of the criteria for success.

Only after comprehensive questioning and detailed documentation of the need, can the hierarchy of the control system be designed. The control system is the means by which the various parts of the handling system are combined to form the whole answer to the problem (ie it is the means of integration). The interfaces between equipments are the stress points in any system and great attention must be paid to them, at all levels of the hierarchy. The requirements of interfaces are defined and examples are given.

At all stages the need to keep the objective firmly in view is paramount and simplicity of design must pervade every level of the control system.

INTRODUCTION

The system chosen as an example of an Automated Handling System is a high bay warehouse with conveyors and AGVs, involving receipt of goods, storage, bulk retrieval, transport and accumulation of loads, load exchange between equipments, and picking. Thus it is intended to illustrate several handling operations combined into one system, ie an integrated system.

ENVIRONMENT

The start point for combining the various parts into a whole is to establish the environment with which the handling system has to be integrated. The purpose of the warehouse as part of the potential users business must be established first.

Warehousing systems may be required to perform one of a number of roles for a variety of purposes, some of which are:-

a   Curing/Cooling

- An example of this is the curing of resin coated sand moulds prior to pouring metal into those moulds.

b.   Quarantine

- This is the situation where the products of a batch process/delivery are subject to quality control testing and the majority of the batch has to be stored pending satisfactory completion of those tests.

c.   Buffer Stock

- This is the case where two consecutive processes vary either in speed or in continuity of operation. This is the most common reason for Work in Progress stock. Most, if not all accumulating conveyor is installed to overcome variations between actions at the on-load and take-off ends of the conveyor. (The average speed of the processes will match but the individual instances will vary).

d.   Process Isolation

- One of the clearest example of this is where the period of production differs from the period of use as is the case with continuous production but intermittent deliveries or despatches.

e.   Kit Picking

- Where the reason for storage is the aggregation of a variety of components and the principle activity is selecting and grouping those components for a subsequent process.

f.   Distribution

- Where goods are received in relatively large amounts from a number of sources and are despatched in small amounts to a large number of destinations.

Having established the role of the handling scheme, (and the throughput rates required) it is possible to address the question of what should the control system be doing and how can it assist in providing an effective service to the handling operation.

It should be noted that the question deliberately excludes functions which are not germane to the handling system, such as Purchasing or Order Processing, nor in the functions of forecasting, exponential smoothing etc which are involved in Stock Control. This is because the control system is a part of the Handling System and, whilst the System and its controls need to be integrated with the user's business, there is no need for it to infiltrate his business and, indeed, it is a positive disadvantage for it to do so.

PURPOSE & CRITERIA FOR SUCCESS

Having established an environment for the operation and a materials handling scheme which solves the clients problem, then the purpose of the control system and some criteria for success can be developed.

It can be generally stated that the controls in an automated handling and warehousing scheme have the following objective:-

- to direct the movement of goods
- to record the stock held
- to control the handling equipment
- & (often) to report to management

The common criteria for a successful design can also be baldly given as Effectiveness, Reliability and Cost. Cost should include not only the initial costs of the installation but also running costs, costs of expansion and the costs consequent upon breakdown. Reliability is not only the assessment of possibility that an item will fail, and the likely extent of that failure, but also the degree to which the effects of a failure can be limited (flexibility in failure), and the maintenance/recovery tools which will enable rapid restart after a failure, plus the absolute neccessity of 100% security of data. It is obvious that Effectiveness is required but technology misapplied is signally unimpressive and it is essential to fit the solution to the problem. Thus we return to the first law of design - First Define The Objectives.

The Objective

In order to reach a precise definition of the objective we need to ask the question "What is the clients problem and what does he need to resolve it" and to apply this to all levels of the system. The purpose, after all, of this (analysis) stage of the design is to refine the purpose of the handling scheme and to identify the activities which need to occur within it.

The most important word here being "need".

Effectiveness

Most clients can identify the general problem to be solved and the role of the system but many find it hard to define the requirements of the solution. The difficulty is to discover and distinguish between Needs and Wishes, Essentials and Desirables (and, in some cases, Fantasies!). Some of the most common sins to which clients can be tempted are:-

- Automation of the manual system (warts and all)
- Empire Isolation (eg Management Systems bypass)
- Wish lists (I want rather than I need)
- WIBNI's (Wouldn't It Be Nice If .....'s)
- Aren't computers wonderful (you can do anything with software these days)
- Knowledge is Power (ie an unbridled desire for masses of reports)

This is not to say that the expression of desires is to be suppressed, only implementation of their more extreme forms. Indeed the optimum solution to a problem may well cater for some of the desires, since what is expressed as a desire may be part of the need.

This analysis of the functions necessary for a computer control system involves posing an exhaustive (and sometimes exhausting) set of questions. To illustrate the range of areas under which such enquiries would be made, the following are some of the chapter headings requiring analysis:-

1. Physical Stock Records
   - At what point does an item become in stock?

2. Location Records
   - What types of storage equipment are used and how many products in one location.

3. Warehouse Map
   - All places where stock may be held, both storage and transit locations.

4. Input Product Details
   - Extent of information and source (man or machine).

5. Output Requirement
   - Extent of information and source (man or machine).

6. Picking List
   - Summary or order. Whole loads or part picking. Is load consolidation required? (In reverse drop sequence?)

7. Location Selection
   - Speed of access, movement factor, low level picking, cross contamination, common products in Drive In/Block Stack.

8. Product Selection
   - FIFO (with tolerance), quarantine, shelf life, batch control, Part full loads, accessibility.

9. Route Loads to & from Storage
   - Resolve priorities, track loads, reconstitute map

10. Print labels
    - Pallet, Product or Destination. Extent of information and is bar coding required or beneficial.

11. Monitor Equipment Health
    - All data links intact, commands acknowledged and performed to time.

12. Print Reports
    - Contents of the store, Product Details, Throughput Reports, Maintenance Reports, Rack Utilisation.

13. Reorganisation of Storage
    - Movement factor, Age, Condemned, Empties, Preparations for peak activity (In or Out).

14. Management Services
    - Available & Free Stock, Movement Records, Pareto Analysis and ABC Classification, Perpetual Inventory.

and all this and then the equipment demands - AGV's, Conveyors, Cranes, Trucks etc plus the number and types of man-machine interfaces (terminals) required.

The documentation of all of this constitutes Section 2 of the Functional Specification. Section 1 is the Introduction and subsequent sections will define, in user terms, the way in which the required functions will be performed. The Functional Specification is the definition of the Objective and is the single most important document of a project. It involves the user (it's his product and his problem) and the supplier (who has to put the solution into practice). It takes a great deal of effort and takes a long time - typically 3 months, could be longer but unlikely to be shorter.

Reliability

Robustness, Resilience, Flexibility in Failure, are all aspects of reliability which have to be considered in terms of likely cost effectiveness. Other considerations will include Utilisation and Availability factors plus the effects of failure of the system (in whole or in part) on the profitability of the clients business.

Any well designed system will include a margin of safety and the system will be tolerant of minor failures. In a conveyor, for instance, where the availability (or reliability) is "x" then, provided that the ratio of Throughput Required to Throughput Capacity (= Utilisation) is less than "x" then the system can recover, without long term consequences. The failure of a section of the conveyor will cause an interruption in delivery of loads but after repair, the conveyor will be capable of working at capacity throughput until the back log of work is completed and the overall throughput of the system has recovered. The time taken to recover from a fault is given by:

$Tr = (Tf \times U)/(1-U)$

where: Tr = Time to Recover
Tf = Duration of Failure
U = Utilisation Factor

In this case there are no long term consequences of the failure and expensive procedures to prevent, or bypass, the failure would not be cost effective.

However, if the full throughput capacity of the conveyor cannot be employed due to external factors such as the maximum rate of feed from preceding equipment, the recovery will take longer or, in extreme cases, it may be prevented. The more common case is where the conveyor utilisation includes a margin of safety but the process being fed is the limiting factor on throughput. In this case any delay in feeding the process cannot be recovered; throughput will be permanently affected, and profit will have been lost. In this instance, it would certainly be advisable to spend money in order to prevent the failure or to provide an alternative means of feeding the process, but the amount of spend should be balanced against the potential loss of profits.

Cost

This last criteria for success, that of cost (or rather cost effectiveness) is a by-product of isolating need from desire, establishing the appropriate level of reliability/resilience and of keeping to simple solutions. Thus the commercial pressures serve to reinforce the principles of sound design.

HIERARCHY OF CONTROL

In the same way as it is possible to give a generalised purpose of an automated warehousing scheme it is also possible to describe a generalised hierarchy of control. The functional requirements of a particular application will determine the details of a particular control system, whilst maintaining a common structure (figure 1).

The upper level of the management hierarchy is the Administrative Systems Computer used by the Company to assist in operating its business and this is supported by the other levels in the hierarchy viz:-

Warehouse Control Computer
Data Collection & Display Systems
 (Computer Peripherals)
Equipment Controllers (eg Micro's &
 Programmable Logic Controllers
Motors, Actuators & Prime Movers
Sensors, Detectors & Measuring Instruments

INTEGRATION AND INTERFACES

Now that the environment has been established, the objective has been defined, the functions and equipment needs have been developed, it only remains to integrate the parts to form the whole answer to the clients problem.

It is the control system which brings the various pieces of handling equipment together to form a cohesive whole, very much like combining the pieces of an interlocking jig saw puzzle. Thus integration is the fitting of interfaces. These interfaces can also be the cause of system failure and thus require very precise definition.

In all but the simplest mechanical examples, interfaces can be defined in terms of the data which passes across the interface, the medium that is used for its transmission and the protocol employed to ensure reliable data transfer.

Data

This is the first characteristic to be investigated and defined. It will include the information required by both sides of the interface and the information necessary to monitor accurate transmission. The manner in which particular items of data are presented will be determined by considerations of:-

- the nature of the interface,
- the required speed of transmission,
- the ease of interpretation by man and
- the costs of the alternatives

Medium

The physical means by which the data will pass across the interface. Examples include light beams between handling devices, multiple wires from sensors to controllers and single cables as required for local area networks or links between computers. The selection of the medium will take account of:-

- the communication devices ("dumb" or "intelligent", fixed or free ranging)
- the data (nature and amount of data to be transmitted),
- the range and speed of communication necessary,
- the reliability of operation (in what may be a hostile environment) and
- the costs of alternatives.

Protocol

The procedures by which data is passed across the interface, the methods of gaining assurance of success and the measures to be adopted in case of failure, ie The rules of Transmission, Verification and Recovery. The rules of transmission must define:-

- the sequence in which the parties exchange data,
- who speaks first,
- how to signal that a message is finished and
- how to tell that it has been received

Verification includes checks on:-

- the integrity of the medium (eg by cycling of switches)
- the accuracy of transmission (eg by parity checks and check digits)
- the validity of the data (eg by checking consistency with previous data)

In the case of failure then recovery methods can range from, stop and raise an alarm for manual intervention, to automatic procedures involving recovery transmissions and identification of duplicate data.

Example Interfaces

Each of the lines joining the boxes in figure 1, Hierarchy of Control, represents an interface but there are many others which also impinge on the design of the control system. The man-machine interface and how the operators relate to the system is one of the more significant factors in ensuring a successful installation. The other major area is that of load exchange between two pieces of materials handling equipment, eg the transfer of loads from conveyor to crane, crane to conveyor, conveyor to lift, conveyor to AGV, etc.

In three interfaces which have been chosen as examples are a Sizing Gate, where the interface is restricted in scope and only one controller is used, a load exchange between equipments, where at least two controllers are involved, and a broad brush approach to the computer communications which occur at the interface between the materials handling system and the remainder of the users business.

Sizing Gate

The situation is where a pallet passing along a conveyor is to be profile checked to verify that its dimensions left and right and its height do not exceed pre-set limits. The method by which this is to be achieved uses photoelectric sensors to detect the presence of a pallet and, to detect load extensions beyond the dimensional limits.

The data passing here is the condition of the light beams and this information has to be received and interpreted by a PLC. The data then is pallet presence beam broken/unbroken, left hand light beam broken/unbroken, right hand beam broken/unbroken and overheight beam broken/unbroken. Manual interpretation of these states can be established by reference to the neon lamps mounted on the detectors themselves.

The medium here is not only the wire between the devices and the PLC but also the characteristics of the signals (eg Volt free contacts, 5V TTL, 24V DC, 110V AC). In this example volt free contacts were chosen with the power being provided by the PLC.

The protocol rules were that the conditions of the photoelectrics should only be monitored in the presence of a pallet, thus obviating at least some of the problems of spurious data. The verification procedures included a system whereby the power to all the photoelectric transmitters was removed and the resulting condition of the photoelectric receivers was verified to be reset (= no light beam detected). No automatic recovery was possible and manual rectification had to be invoked via an alarm.

Load Exchange between Handling Equipments

The particular example chosen is the situation where a pallet load is transferred from an AGV to a conveyor. Transmissions between the two equipments is by photoelectric transmitters and receivers; the AGV carries one transmitter (A) and the conveyor (the load receiving equipment) has two transmitters (B & C). The starting condition is with all transmitters off.

To effect the load exchange then the AGV arrives opposite the conveyor and switches on its transmitter to indicate "All my systems are healthy, I have a load to exchange, are you ready to receive?" In response the conveyor transmitters B and C are switched on to state "All my systems are healthy, I am ready, let's do it." The appropriate motors will then be started, the load will transfer from one device to the other and the progress of the load will be monitored by both equipments. The fact that the load has left the AGV will be recognised on board the vehicle and the fact that the load is fully onto the conveyor will be detected by sensors on the conveyor. At this point the conveyor transmitter B is switched off to indicate "Thank you, load transfer is complete, all is well." The AGV will acknowledge receipt of this signal by switching off its transmitter. When the conveyor controller detects this at the end of the cycle, it responds by switching off its transmitters. This state is recognised by the AGV as signalling a healthy reset condition and it is then free to undertake further work.

The normal load exchange sequence is shown in table 1 and appears pretty basic. However, it does illustrate some important characteristics:-

a. All steady state conditions result in unambiguous diagnosis of action required.

b. Interpretation of the signals is not dependent on the sequence of reaching the condition.

c. All the signals must be seen to change state before motors start, ie the integrity of the media is verified before action is taken.

d. Fault conditions (also shown in Table 1) are specific and identify the item at fault.

e. All fault conditions call for the same action = Stop all motors and raise an alarm.

Equipment	AGV	Conveyor		Binary
Transmitter	A	B	C	ABC
Normal sequence:-				
1. Initial Conditions	OFF	OFF	OFF	000
2. AGV loaded & ready	ON			100
3. Start load transfer		ON	ON	111
4. Stop load transfer		OFF		101
5. AGV acknowledgement	OFF			001
6. Healthy Reset			OFF	000
Fault conditions:-				
1. AGV Unhealthy	OFF	ON	ON	011
2. Conveyor fault	ON	ON	OFF	110
3. Both faulty	OFF	ON	OFF	010

Table 1 Signal Conditions for load exchange

Thus the data is the status of the equipments and the commands to start and stop the load exchange. The medium is the infra-red light between the photoelectric transmitters and receivers and the protocol is the transmission, verification and alarm procedures described above.

Communications between the Control and Management Computers

Where there are significant amounts of data to be exchanged between the computers (as is usually the case) then the data should be passed from a file on one computer to a file on the other using an appropriate protocol (or device emulator). The use of a file to file transfer of data gives each computer a degree of operational isolation from the other - if one breaks down (or the link fails), then the working computer(s) can continue to work on the information already received, and it can store the data to be sent in its own transmission file pending repair of the faulty equipment.

The usual form of transmission will be a synchronous serial link across a fixed wire but use could be made of the facilities of Mercury/British Telecom if the administrative computer is on a remote site, in which case the media used may include fibre optics and/or microwaves in addition to fixed wiring.

There are several "standard" protocols and emulators which can be readily purchased and this is more cost effective than writing one anew for each application. A good example of this is IBM 2780 (though this is now being superceded). It should be noted however that one manufacturers implementation may not be identical to another and difficulties can arise. For instance, there are speed limitations when DEC and HP communicate using 2780 emulation, (although both would claim there is no such limitation when communicating with IBM).

CONCLUSION

There are many aspects of integration which need to be considered when designing a handling scheme, and the control system not only has the role of integrator but must itself be designed such that the various parts are combined to make one whole system.

The Last Word - Simplicity

This is not to be regarded as a constraint upon the objectives of a system, since some projects can appear to be very ambitious, but it must be regarded as the by-word for all phases of the design. KISS, (Keep It Simple, Stupid) should be constantly born in mind. The aim is for "Plain Vanilla Flavour", no "Bells and Whistles", no "decoration", no "WIBNI's", no "Extras" and nothing "clever". The cleverness of the designer should be exhibited in producing simple solutions. Elegance of design is not revealed by a system's complexity but by its simplicity.

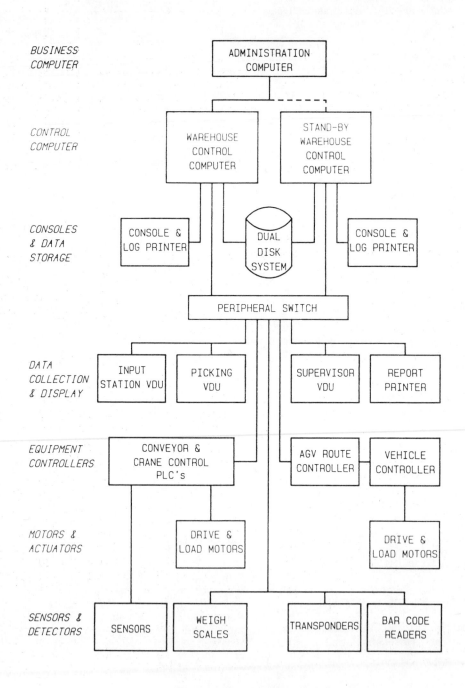

Fig 1 Hierarchy of control

Use of intelligent pallets in the Personal System 2 test process

C G McDOWELL, BSc
IBM UK Limited, Greenock, Renfrewshire, Scotland

SYNOPSIS A Personal Computer System Test process has been developed that makes use of intelligent pallets both for process control and system test functions. The project offers positive indications that a system based on decentralisation of process control through use of autonomous intelligent pallets is capable of delivering improvements in process control and process reliability.

1. INTRODUCTION

Personal Computer system test processes involve several steps. The product is first Screened to remove assembly errors. The product is then powered up and operated for a four hour period. This RUNIN operation is performed to provide a measure of the reliability of the product. At the completion of this process step the unit is transported to a Final Verify process step where the results of the RUNIN operation are checked. Failing units are routed to a rework operation before re-entering the process.

The multi step operation is typical of many manufacturing processes and to ensure a high level of integrity of the product leaving the production line a process tracking system is required. This generally takes the form of a centralised data collection system with all process steps reporting directly to it. A complex manufacturing subsystem generally evolves with a tightly knit combination of Screen Capture stations, Runin racks, Final Verify Capture stations and Rework stations. The whole process is interdependent and failures in the process control system are catastrophic to the total process. In an effort to eliminate the need for an overall process control system an intelligent pallet carrying the process control information was developed in Greenock. This pallet takes the concept of intelligent tags further and provides product test intelligence as well as process tracking intelligence.

The product test interface to the Personal System product involves some fifty wired connections. No previously developed automated process had this complexity of test interface. Known long term reliability problems with pallet connecting pins required the number of hard wired connections to be minimised. These would normally be provided by wiring the product via a cable set to a set of connectors on the pallet which would interface the product to the process test equipment. These connections will detract from the overall reliability of the system. In the developed system the product test interconnections are multiplexed via the pallet intelligence to communicate through an infra red link to the test equipment. All process steps communicate to the pallet via the infra-red link and a process tracking facility is provided by a non volatile memory read and written via the infra red link.

2.0 THE INTELLIGENT PALLET

The pallet as implemented for the Personal System 2 Test process has the following elements (FIGURE 1)

o 8749 control microprocessor (2.5uS)

o 1K byte/sec Infra Red control link

o 128 bytes serially read/write EEPROM

o Universal Personal System test
   - Line Power ON/OFF control port
   - Keyboard simulation Port
   - Parallel Port attachment
   - Video Port Test logic
       Analog level check
       Pixel transition count check

The Infra Red link uses industry standard television type components and encodes data using a scheme of Pulse Position Modulation.

A PULSE is 6 microprocessor clock cycles
A ONE is a pulse gap of 13 clock cycles
A ZERO is a pulse gap of 19 clock cycles
as part of the transmission there is an interbyte gap of 35 clock cycles to accomodate processing delays.

Two FF characters are sent as a preamble to each message to allow the Automatic Gain Control system of the Infra Red Link to settle. This is followed by two message header characters STX($02) and $00 then the byte count of the message , the command byte followed by two bytes addressing the individual pallet.

   FF FF 02 00 bb cc pp pp mm mm mm m..

 bb is the message byte count
 cc is the pallet command
 pp is the pallet identification number
 mm is the message data stream

The possible pallet command data streams and their responses are defined in FIGURE 2.

Each transmission requires a response. Some responses indicate that the function has been performed correctly. Other time dependent functions merely that the message has been received correctly and will be acted upon.

As an example of this scheme-

WRITE to EEPROM location 23 data as $55AA. The serial EEPROM used is organised two bytes per location.

Transmission -

    FF FF 02 00 06 32 FF FF 23 55 AA

$32 is the write EEPROM command
$23,$55 and $AA are the data indicating
    write of $55AA to location $23

Response -

   FF FF 02 00 01 06

indication operation was successful. $06 is ACK, $07 is sent as unsuccessful NACK respose.

An identical logic card to the one housed in the pallet is used to perform the control communication function. It is attached to a Personal Computer using the parallel port interface. The command to be sent and the responses from the intelligent pallet are communicated over this link. Only the 8749 microprocessor is reprogrammed to handle this function. Pallet commands are defined as PASCAL FUNCTIONS and these are used to control the operation of the intelligent pallet.

The pallet continuously polls the product keyboard and parallel interfaces plus the Infra Red Link for activity. All of the pallet functions that can be initiated via the Infra Red link can also be controlled from the product via the parallel port. Messages are sent in the same format to the pallet but without the FFFF preamble. All parallel port messages sent with a STX($02) header are handled and responded to as though it had been received via the infra red link. The product can write test process results to the EEPROM for later readout via the controlling Infra Red link. The product can power itself down at the completion of test or it can initiate a Power Cycling sequence via the PowerCYCLE command.

3.0 DATA COLLECTION SYSTEM

A PC Local Area Network connects all test controllers in the system. Each controller in the system collects its own performance data locally. The operation of the whole system is not dependent on the operation of the data collection system which merely processes data collected in the distributed databases for reporting purposes. Routines running in a system attached to the network periodically poll the distributed data base for activity and provide system performance statistics.

The need for 100 per cent availability of the Data Collection System is removed since the operation of the process is no longer dependent on it.

- the product test failure code is
  stored in the pallet EEPROM

- Information on the path of the product
  through the test system is stored in
  the pallet EEPROM

- all data is collected locally at the
  test stations.

4.0 CRANE CONTROL SYSTEM

During the Personal System 2 testing the bulk of the process occurs during RUNIN. This is an operation where the product is powered up and cycles on a predefined set of diagnostic tests logging test results onto floppy disk during the process. Units are stored for this process step in an Automatic Storage Retrieval System (ASRS). The ASRS crane is controlled directly by an Allen Bradley PLC operating under control of an IBM Industrial Personal Computer (PC). Both the PLC and the PC are mounted on the crane. The Industrial PC on the crane communicates directly to a further ground based PC via a serial data link (Figure 3).

The Crane based Industrial PC controls the following functions.

- Scheduling of pallets in and out of the rack.
- Crane error recovery.
- Communicates to pallets in the racks via an Infra Red Link mounted on the crane

The Ground based PC controls the following functions.

- Maps crane based PC activity
- Allows entry of manual commands so that maintenance functions can be performed.

In operation a pallet is picked up at the crane Infeed and placed in the nearest empty location in the racks. The Crane then checks the pallet EEPROM for the product serial number and the results of the capture station safety test. If the result indicates a failure the pallet will be sent to the Reject queue on its way to the Rework operation. If the unit is good it can be powered up via the IR Link and testing started. At the completion of testing the test diagnostics will write the test results to the pallet EEPROM and command power to be removed from the product. At the end of the RUNIN period the crane will interrogate the result. The product will be sent to the ship/pack operation if the result is good or to the Rework operation if a failure is indicated.

5.0 PROCESS OPERATION

The manufacturing test process implemented based on these elements has the form defined in Figure 4.

Each Personal Computer product has a bar coded label with serial number and model information. This is scanned by a laser reader and the contents written to the pallet EEPROM. If the previous serial number is not equal to the scanned one then it is assumed that this unit is new to the process and all data in the pallet is cleared. If the bar code fails to read and the unit was previously at a rework operation then it is assumed that the serial number in the pallet is that for the product currently on the pallet. If the product has been to a rework operation then testing which has been performed there is not repeated. The product RUNIN time is determined and written to the pallet. Mandatory safety tests are then performed and the results recorded. The product is then powered up and initial Pre RUNIN testing performed and the result recorded. The results of all testing performed is also logged in a database local to the test station.

The unit then enters the ASRS for the RUNIN operation. After the unit is plugged to a rack position RUNIN time and serial number data are read by the crane and logged to the crane RUNIN map. The product is then powered up and the floppy disk based diagnostics are executed for the RUNIN period. At the completion of this time the test results are written to the pallet. The crane control code then identifies the unit as having completed RUNIN and depending on the actual form of testing implemented the product can be dispositioned (PASS or FAIL) directly from the racks or sent to a final test operation where a visual inspection is performed on the product and the test results read from the pallet.

6. CONCLUSIONS

Process control using information stored at the work object simplifies the design and development of the total process.

o Improved process control flexibility
o Reduced probability of process walkthroughs.
o Reduction of process steps
o Changes at individual process steps can be made more in isolation from the overall process.
o Zero dependance on availability of a centralized process control system
o Immunity of process to catastrophic failure modes.

The system does however have its own drawbacks. A system such as the one described which combines test and intelligent pallet functions adds a level of complexity to the equipment to be handled by production line personnel. There is a level of damage/failure in the equipment which will occur regardless of the goodness of the process installed. It requires manufacturing line discipline to manage this situation and to arrange for testing/repair of faulty pallets as they occur.

The problems we have seen in this respect do not relate to the tracking functions of the pallet but to the imbedded test functions and the complex product interconnections. Though catastrophic failures do not occur in this type of process it is apparent that some of the process reliability benefits of this system must be offset by consideration of a degradation in process capacity when faulty process hardware is allowed to accumulate.

Acknowledgements.
The author acknowledges the work of D Barclay, E Keane, A Reid, G Sweeney in the development of the basic hardware and software described.

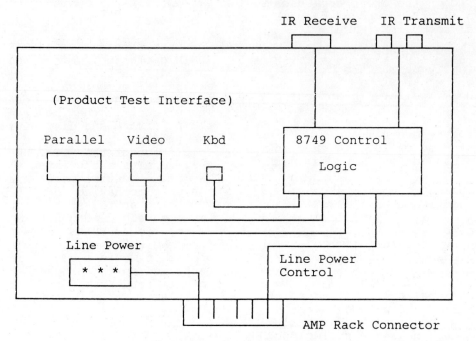

Fig 1   Content of intelligent pallet

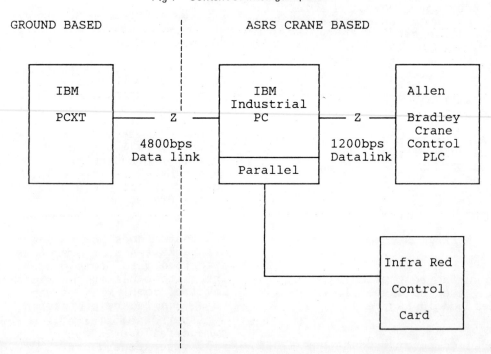

Fig 3   Crane control subsystem

	Command	Command data stream
16H	Read from parallel port using protocol	none Reply - ACK plus port data or NACK
18H	Write to parallel port using protocol	Parallel port data byte Reply - ACK or NACK
22H	Power up Command and send startup byte.	Keyboard start up byte Reply - ACK only if received
28H	Write to Keyboard port	Keyboard port data byte Reply - ACK or NACK
30H	Read from EEPROM	EEPROM Data address byte Reply - ACK plus MSB+LSB EEPROM data or NACK
32H	Write to EEPROM	EEPROM Data address byte MSB of data LSB of data Reply - ACK or NACK
34H	Power Cycle command	Keyboard start up byte Power Down time byte(seconds) Reply - ACK only if received
36H	Read from RAM	RAM Data address byte Reply - ACK plus RAM data or NACK
38H	Write to RAM	RAM Data address byte Data to write byte Reply - ACK or NACK

40H	Block data transfer	Command subtypes byte
		
		  -11H infra red wrap
		  -22H parallel port write
		  -33H keyboard port write
		+ data to write or wrap
		
		Reply - ACK or NACK

Fig 2 Intelligent pallet commands

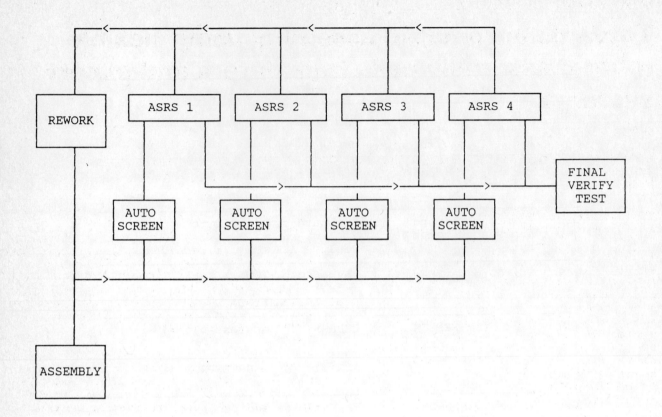

Fig 4 Intelligent pallet-based personal computer test process

C86/88

Towards automated assembly using flexible robotic assembly cells, transputers and expert systems

M TAIO, BSc
Institute of Intelligent Machines, Academia Sinica, Peoples Republic of China
R SEALS, BSc, PhD, AMIEE, **R GILL**, BSc, PhD, MIQA, **R RUOCCO**, BSc, CEng, MIEE, MSPIE and
A S WHITE, BSc(Eng), MSc, CEng, FRAS, MRAeS
Middlesex Automation and Robotics Centre, Middlesex Polytechnic, London

SYNOPSIS The paper describes the progress to date at Middlesex Polytechnic to build a flexible robotic assembly cell for small batches of different products.

1 INTRODUCTION

Automated assembly has become one of the most active areas in robot application due to predicted improvements in quality and productivity. Already some form of automated assembly is being used in 50% of radio and television manufacturing, 30% of white goods and in automobile production (1). However, for small batch sizes automated assembly has not yet made any great impact although its potential has been widely recognised.

An analysis of the process economics (2,3) has shown that production of fixtures for parts feeding, orienting and holding, and system programming are major obstacles to the implementation of automated assembly for small batch sizes.

1.1 Flexible robotic assembly cell

A concept of flexible robotic assembly cell (FRAC) is presented in this paper which seeks to overcome the limitations outlined. A FRAC is a general purpose assembly cell which is being developed from the ideas that follow.

The FRAC should be able to assemble several kinds of products continuously without re-tooling or reprogramming in order to achieve the scale economies of continuous production from several small batches of different products. General purpose part feeding, orienting, holding devices and sensors are used to reduce tooling costs for products which are complex to assemble and have short production life cycles.

The FRAC is being designed so that the specifications of the product to be assembled are entered into the control computer of the FRAC via a computer aided design and manufacturing system using a classification and coding system which not only defines the component parts but also additional instructions on the assembly procedures to be used. This information is analysed by the control system, using one or several expert systems to determine what machines, fixtures and tooling are required, the sequence for assembly, and the presentation of components or sub-assemblies to the FRAC from the external environment. This is then compared to an existing FRAC structure which indicates the number and types of robot arms, tools, parts feeding capabilities and sensory systems required. Expert systems are ideally suited for this matching exercise, and if the current FRAC configuration cannot cope with the assembly task the minimum changes necessary will be indicated.

A FRAC is not a static collection of assembly machines but a flexible arrangement of processes which can be altered if required. The control system will be flexible, alterable and should require minimal reprogramming for each new FRAC arrangement. The FRAC comprises of three functional blocks; action machines (e.g. robots, materials processors or both), sensors and the control system.

1.2 The control system

The configuration and operation of the FRAC is such that it demands a control system which is easily alterable in order to control a wide variety of action machines simultaneously in real time, be able to communicate with other computer systems and to interpret other languages using protocols such as the manufacturing automation protocol (MAP).

This is apparently very difficult to achieve but the use of transputers (4) enables the majority of the requirements to be fulfilled. The transputer can be considered as a black box which is capable of executing processes concurrently and can be connected to other transputers simply using built-in serial communications links.

Occam is the language used to program the transputer which has been specifically written to ease the task of developing programs which execute concurrently, and is independent of the number of transputers connected together. Increasing the number of transputers increases the speed of execution of the control program. An example of a simple concurrent program coded in Occam follows and figure 1 illustrates how the processes can be executed on a single or multiple transputer system. The inter process communication is automatically up-dated without making any changes in the program.

Concurrent program

```
PAR
  Assemble-sub-component-1)   All four sub com-
  Assemble-sub-component-2)   ponents will be
  Assemble-sub-component-3)   assembled simultan-
  Assemble-sub-component-4)   eously.
```

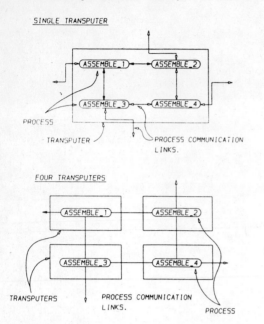

Fig 1   Process execution on a single or multiple transputer system

A FRAC which consists of a small number of action machines could be controlled by a small number of transputers; however, more complex FRAC configurations may require additional transputers to execute in the required time-scale. Due to the high operating speeds of the transputer allowance for the individual dynamics of the action machines can be taken into account to increase the dynamic accuracy of the assembly process. The control system will consist of expert systems which match the resources of the FRAC against those required for a specific assembly task (fig. 2), and will then direct the control commands for the action machines using sensory feedback.

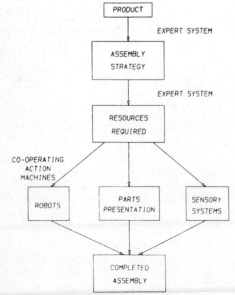

Fig 2   Control system

Each action machine will require a software driver capable of communicating with all other action machines in the FRAC. The communication will be provided by the inherent serial links between the transputers; this obviates the need for a special communication system for interaction machine exchange (fig.3). The software could be written in Occam and executed on a single transputer, or alternatively, since each of the action machines will already have a controlling computer and appropriate software, they will be linked to the main Occam program, which will execute on several transputers.

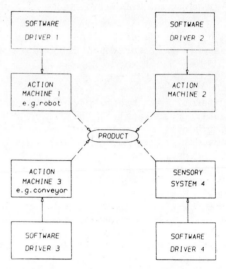

Fig 3   Communication links

The expert systems could be written in Occam; however, it is more likely that another language which can be executed on the transputer will be used.

1.3 Expert Systems

Expert systems are ideal for assisting assembly programs when used to match resources against demand. At present an expert system (which forms the basis of the expert super system) is being developed for use with the FRAC and to date it is capable of monitoring components from a variety of assemblies presented in a random order. The system then identifies whenever sufficient components have been received to put together a particular assembly (fig. 4). Although this is not a particularly difficult task and could be easily achieved using other techniques, the expert system is designed to be a fore-runner of a suite of programs which will assist assembly throughout the FRAC. The expert system has been developed in Prolog and will either be recoded in Occam or used directly by the transputer via a Prolog compiler.

1.4 Sensor Suite

The main feature of the FRAC is its flexibility of operation, namely its capacity for carrying out simple but different tasks on a given number

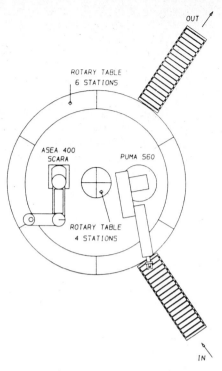

Fig 4 Flexible robotic assembly cell

of different parts to be assembled. A complex series of operations can then be executed by connecting several FRACs together in a combination of serial and parallel configurations. Each FRAC therefore requires a sensor system capable of identifying and orienting an assembly part from a batch of randomly supplied items (thus reducing the need for dedicated and expensive feeders), as well as providing information about the parts' interaction throughout the assembly operation.

To attain the required accuracy whilst maintaining flexibility of operation such a sensor system requires a 3-D vision sensor, a tactile sensor and a force sensor. The 3-D vision sensor consists of two orthogonally mounted cameras which provide a top and side view of the part to be assembled. The tactile sensors are mounted within the robot gripper to provide a pressure image of the part being handled and the force sensors are mounted within the robot wrist to provide 6-D data of any dynamic robot operation such as component insertion. Such a multi-sensory system is clearly advantageous in terms of operational flexibility but increases the amount of sensory data beyond the processing power of most serial control computers. For example two images with 256x256 resolution and 8 grey level bits, provided by the 3-D vision sensor, yield 128K bytes of data per 'frame' whilst the two 16x16 tactile sensor pads add another 512 bytes to the computational load for every sampling period. An expert system based, robot control system capable of parallel processing is being developed at Middlesex Polytechnic with the aim of extracting maximum information from such multisensory data to allow task interactions at a rate suitable for real time FRAC operation based on a 4 transputer network. The use of a transputer network is expected to provide full error recovery when used in conjunction with an expert system (which stores 'world' as well as assembly parts models) and the multisensor feedback system. The parallel control algorithm will continuously correlate the measured sensory data with the expected information provided by the models within the expert system. Rapid convergence of the decision loops within the algorithm is possible even in the absence of complete sensory data, for example, when using images with partly occluded objects.

The state of readiness of the component parts of the sensor suite is that the vision cameras and their associated software are fully operational, the force sensor has been made and is functional, the only incomplete part of the system is the tactile sensor.

2 THE FRAC PILOT SCHEME

Any system for automated production must sooner or later be evaluated against competing schemes to ascertain its efficacy. Control using a transputer must be compared with traditional sequential control in order to obtain a true measurement of its worth.

To evaluate the schemes an assembly task was chosen which fitted in with the machines available at Middlesex rather than design new machines to assemble the parts. The assembly of a marine toilet (fig. 5) is a task which requires considerable flexibility from the action machines so that variations from the basic model can be produced to fulfil the market demand. The tools required have to cope with parts with outside dimensions which are not as precise as metallic components, and the gripper jaws must be compliant so that no permanent damage is inflicted on the parts.

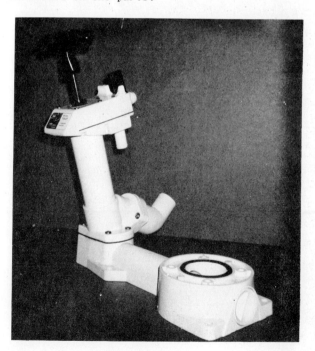

Fig 5 Toilet assembly

A number of factors affected the choice of assembly for the pilot scheme:

a) weight of the components,
b) the shortest time available between component changes,
c) the number of different configurations within the FRAC at any one time,
d) the variety of finished parts reaching the stations in a random sequence, and

e) limitation of the control system.

Some conclusions (which may be changed later) were reached,

a) the components should be handled by robots,
b) the delivery of formed parts would be the input to the control system,
c) no storage at the assembly station,
d) finished assemblies to be dispatched immediately to remove the need to store parts (to help just in time procedures),
e) forward planning using data from the sales computer, allows decisions of exact progress of components to time their arrival at the FRAC.

The decisions (a) to (e) were made to ensure that the maximum flexibility is incorporated into every station within the FRAC.

2.1 Action Machines

The FRAC consists of two robots, a selective compliance assembly robot (SCARA)-ASEA 400, and a PUMA 560, two rotary tables (fig. 4), component orientators and translators, and two gripper banks, one for each robot.

The SCARA robot which is ideally suited for vertical stacking operations has four degrees of freedom and can handle objects up to 4kg. Its main task is to locate the components into the assembly using suitable grippers selected from the gripper bank. The Puma robot is used for component transportation between the conveyors and the assembly tables, orientation, and for secondary operations such as the driving home of screws and manipulating glue applicators. Both robots have access to the two rotary tables, which have a total of ten work-stations. Simulation packages (SIMAN, CATIA) are being used to locate possible robot collisions, but this task will be conducted by the expert system when the FRAC is fully operational. Each station on the rotary tables has a jaw mechanism for gripping the sub-assembly or other components, an orientation unit for rotation in two axes, a translator for horizontal motion and a scissor linked platform for height adjustment. The translator, orientation and scissor units are powered by locally mounted DC servomotors.

At present most grippers are application dependent. However, it is intended that a family of gripper fingers will be designed for a range of end effectors incorporating the sensors mentioned earlier. If any partially assembled products fail inspection (conducted within the cell by an on line automated vision system which checks the presence and location of all components), further processing of the particular assembly is stopped and after a completed cycle where no further work is done on the assembly it is redirected via the output conveyor to a parts reclamation and rectification station (manual rework station). If a fault recurs persistently the FRAC is stopped and the error recovery system is interrogated. This can also be interrogated whilst the FRAC is in full swing.

2.2 Modular System

A FRAC in a flexible assembly line could be isolated or removed completely/replaced by another FRAC on a modular basis for 100% inspection, analysis and for maintenance. New modules could be added to extend the lines with minimal disruption of production.

2.3 Parts Feeding

Each FRAC could have its own automated warehouse (linked via the input conveyor). The parts are picked from the conveyor fed by the automated warehouse in the correct sequence. In the case of an assembly line numerous automated warehouses could be linked together.

2.4 Pilot Cell - Product Choice

The product chosen was not designed for automated assembly (figs. 5 and 6), but for maintainability (especially accessibility) in cramped conditions.

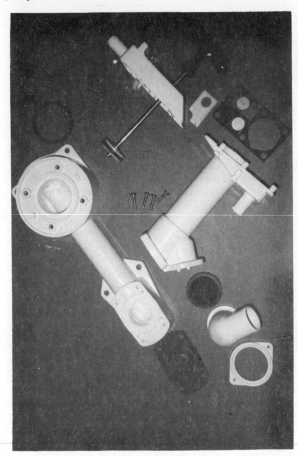

Fig 6   Toilet parts

The maintenance aspect and design for automated assembly have one fundamental thing in common - the majority of the component fastening operations are made simpler, and the robot can gain access with relative ease. Although the number of parts can be reduced, the corresponding manufacturing costs would increase due to the materials processing operations involved (tooling and other costs). One major benefit in assembling this product was the inherent compliance of the parent material.

Initially assembly operations have been centred around the major components - that is, the base and the pump. In the near future all the sub-assembly operations will also be attempted using the basic FRAC configuration.

2.5 Assembly sequence

Only the major operations are shown for clarity (check pick up and replacement are omitted)

Table 1

No	Operation	End effector	Part	Sensor
1	arrival at point 1 on the conveyor		base of pump	light beam to stop the track vision to detect orientation and type
2	pick up by PUMA and insert into rotator	parallel jaw gripper	"	vision sensor enables close control of part pick up
3	jaws of rotator close			force feed back on jaws
4	pick up 'O' ring and MOVE	3 jaw compliant with outer ring	'O' ring seal	vision
5	locate 'O' ring in groove and insert		"	
6	pick up valve pad 1 SCARA and MOVE	vacuum gripper	valve pad 1	vision sensor + pressure sensor
7	locate on base and MOVE		"	
8	pick up pump body with PUMA & MOVE	parallel jaw gripper	pump body	vision
9	locate on valve housing HOLD			"
10	pick up screws with SCARA & MOVE	screw holder	screws size 1	vision
11	insert into predrilled holes and rotate		"	
12	repeat 10 & 11 3 times		"	
13	rotate rotator chuck			LVDT
14	pick up valve pad with SCARA & MOVE	vacuum gripper	valve pad 2	vision and pressure sensor
15	locate pad and MOVE			
16	pick up plunger assembly & MOVE	parallel jaw gripper	plunger assembly	vision + tactile sensor
17	locate plunger			vision
18	pick up screw MOVE	screw holder	screw size 2	"
19	locate in housing and rotate			"
20	repeat 18 & 19 5 times			
21	rotate pump body in rotator			LVDT
22	pick up exit pipe in PUMA & MOVE		exit pipe	tactile sensor
23	locate on housing and HOLD			vision and tactile
24	pick up screws	screw holder	screw size 1	vision
25	insert and rotate			
26	repeat 24 & 25			
27	grasp pump body in PUMA	parallel jaw gripper	pump body	tactile sensor
28	release rotatorjaws			
29	move part to table			vision
30	locate body on bowl		+ tactile	
31	pick up screw SCARA & MOVE	screw holder	screw size 3	vision
32	locate and rotate			
33	repeat 29 & 30 3 times			
34	bowl + pump transferred to conveyor			

2.6 Simulation

Any system of production in which the possibility of disaster is significant will benefit from simulation. In the case of the pilot scheme two types of simulation are being used, one to ascertain the kinematic restraints of the cell using CATIA robotic software on 5080 workstation running on an IBM4381 mainframe computer. This software allows the user to input a database for the machine in question, say, a robot and its motion is then generated by the CATIA software. Comparison between the simulation and measurements of the true robot motion have shown data inaccuracies in the robot specifications. Further work is being undertaken to rectify these faults in the VAL robot language. The time scheduling of the production sequence is being made with SIMAN a discrete simulation language.

In order to model the dynamic performance of the control system a full description of the cell dynamics is being undertaken with ACSL, a continuous simulation package and Control C for control system design. The prime object here is to develop a facility to check the cell performance before production changes are made.

A Petri-net-based logic controller (PNLC) developed at Middlesex has been used to emulate and simulate the interaction of the logic program with the controlled hardware prior to installation(5). This simulation package can also be used to model complete processes thus making the PNLC a useful tool for the development of automated manufacturing systems.

3 CONCLUDING REMARKS

The development which is outlined in this paper is of a flexible Robotic Assembly Cell. This cell has a number of reprogrammable elements under the supervision of a transputer network using several expert systems to decide between conflicting data from a sensor suite and the production program for a given assembly task. Many advantages accrue from the comprehensive nature of the FRAC which are mainly:

    enhanced flexibility,
    speed of decision making and data processing,
    shorter hardware development time;

while the prime disadvantages are:

    increased software and concept development time,
    initial cost of cell.

The last comment should be seen as a two-edged sword in that new products can be dealt with in much less time than the cell development and for little or no extra cost. Initial development at Middlesex has shown that all the sensors can be made to work effectively in a harsh environment; the expert system can distinguish between conflicting sensor data and provided the initial choice of products for a given cell is made with regard to the performance of the action machines, the system will work effectively. Although the FRAC at Middlesex is not complete most of the components of the system are tested and await system 'tuning'.

REFERENCES

(1) MILLING, A. New manufacturing concepts - the plant engineers perspective. The World Yearbook of Robotics Research and Development.
pub. Kogan Page Ltd. 1986 pp 85-93.

(2) BOOTHROYD, G. Economics of general purpose assembly robots. IFS(Publications) Ltd. 1985, pp. 335-345.

(3) MILLER, J. Robot assembly system design, assembly cost, and manufacturing viability.
Robotic Assembly, IFS(Publications)Ltd. 1985 pp 319-334.

(4) Production Information: The Transputer Family Inmos 1986.

(5) TIZZARD, G A., GILL, R., A Petri-Net Based Programmable Logic Controller, Emulator and Simulator. Proc. of the 4th Conf. in Manuf.Technology - Ireland 1987.

C87/88

A comparison of the cost characteristics of modern parts storage and handling systems

K FIRTH, FIMM, **C TURNER**, BA, MSc and **H PAVELEY**, BA, MSc
NMHC Consulting Group Limited, Cranfield Institute of Technology, Cranfield, Bedfordshire

INTRODUCTION

Many manufacturing and distribution processes have a need to store small components, often in ranges numbering 10,000 different items and more. Keeping an account of such ranges in terms of how many are held, where they are stored, when they should be issued, and how much loss is occuring due to bad practice has long been a weakness. The advent of CIM and computerised distribution systems has given us the opportunity of more sophisticated control and correspondingly sophisticated storage and handling systems. Each of these systems has its own attributes and its protagonists and each is in heavy competition with the other - the big question which arises is how does each compare from the viewpoint of cost and operation. This paper is an attempt to unravel the morass of information - sometimes misleading - which accompanies the introduction of new ideas. It does not set out to look at every aspect of small parts storage and retrieval just the main contenders in the field at this point in time.

It is first necessary to examine the main constituents of modern small parts storage and handling. Which are:

(i) Picker to goods.

(ii) Goods to picker.

(i) <u>Picker to Goods</u>

This is the more traditional method of storage, that is where the stock is arranged on shelves or racking and the picker, on foot, or on a vehicle goes to the storage location to access the required stock. Within this case are included all the totally manual systems, but of more interest are the semi-automatic systems, i.e. fixed-path or free-path order pickers. These vehicles travel along the racking, either under manual or automatic control, to the required picking location. They will be described more fully in the respective comparative sections.

(ii) <u>Goods to Picker</u>

These systems have a higher level of mechanical sophistication, in general terms the picker stays in one position, whilst the selected location moves to the picker. They can be sub-divided into horizontal carousels, vertical carousels, and mini-load (AS/RS). The characteristics of each system will also be discussed in their individual cases.

COMPARISONS

The method which has been chosen for this study is to compare each system under review taking into account the major variables which can effect cost. These are building occupancy, level of stock, level of throughput, shifts operated, capital cost of equipment, labour required, maintenance and central costs - all of which interact in ways which are indeterminable by subjective assessment. To do this we have set a basic stockholding unit of a tote-pan which is 0.01 m^3 but it should be appreciated that our technique is amenable to modification in this respect.

The stockholding figures chosen are as follows:

1,000 tote bins

2,500 tote bins

5,000 tote bins

10,000 tote bins

20,000 tote bins.

The throughput figures are in terms of individual accessions (combined input and output) and the levels used are:

100 picks/day or 26,000 picks/year

250 picks/day or 65,000 picks/year

500 picks/day or 130,000 picks/year

1,000 picks/day or 260,000 picks/year

5,000 picks/day or 1,300,000 picks/year

No system is feasible for every situation considered, this is not however a surprising result, as each system has been developed to satisfy a particular range of applications. Although cost is a very useful basis by which to compare systems it is not the only criteria to consider as some systems have inherent advantages which for specific applications may outweigh their higher cost.

COSTS

As the varying systems utilised different heights, the building cost is related to the specific height in each case. By referring to the graph (Appendix 1) the cost per square metre for any height of building can be ascertained. The building cost so obtained includes current UK prices of new buildings, local rates, heating, lighting, ventilation, building maintenance, cleaning etc. Racking costs include the purchase of tote boxes.

Because of the need for regular layout some systems provide more than the specified number of locations, however, only the cost for the specific number of tote boxes has been included. Some systems do not include racking costs as a separate item because storage and retrieval equipment is integral to the system.

Machine cost for 'picker to goods' systems account for the order picking vehicle, whereas in 'goods to picker' systems the machine costs account for all the storage and retrieval equipment.

All the above capital costs have been rentalised on a per annum basis using DCF at 11 per cent over the estimated equipment life.

Labour costs include 'on-costs' and also take into account shift premiums for 2 and 3 shift operations.

Equipment maintenance is an annual cost comprising a percentage of the capital cost, and varying depending on the type of equipment and the number of shifts over which it is used.

Control costs are also included, which vary according to the different systems.

The individual costs have then been totalled to give an annual operating cost, and then divided by the throughput, to arrive at the unit cost of throughput.

TYPE 1 - PICKER TO GOODS SYSTEMS

1.1 Free-path order picker

The order picking vehicle operates in aisles between racking. If the aisles are wide enough, the machines can pass one another. Alternatively, the aisles can be kept narrow to improve space utilisation, in which case zone picking should be used otherwise queueing becomes a problem.

Recent advances have improved this system considerably. Portable terminals introduce the possibility of paperless picking and, added to bar-code techniques the accuracy of picking can be improved.

A further advance is the wire-in-floor technique, while in the aisle the vehicle is guided thereby reducing the incidence of damage caused by poor driving technique.

The picking rate is approximately 55 picks per hour for the purpose of this study.

Table 1 - Costs of free-path order picker system

1.2 Fixed-path order picker

Fixed-path systems are economical in high buildings, and they are most effective at heights in excess of 6 m. In this case a rack height of 12 m has been used. They are most useful in high holding, high throughput situations. (Throughput of 100 picks per hour are reasonable.)

Although it is possible to move the order-pickers from one aisle to another, this involves either a specialised order-picker, or a transfer car, hence during this study we have assumed one machine, one aisle.

As with free-path order-pickers 2-way radio communication and on-board computing can be used to enhance their effectiveness.

Table 2 - Costs of fixed-path order picker systems

TYPE 2 - GOODS TO PICKER SYSTEMS

2.1 Vertical carousels

A vertical carousel consists of a series of shelves linked by a chain drive, the whole system is enclosed with an opening for the picking face. Being enclosed, a vertical carousel protects items from accidental damage, dirt, pilferage, etc. Therefore making them very suitable for shop-floor activity. They also offer advantages in terms of fire protection, allowing for instance, the localised use of halon gas for water sensitive items such as electronic components.

They can range in height from under 2 m to over 7 m, with weight capacity ranging from 75 kg - 250 kg per carrier. Picking rates can be expected to be 100 - 110 picks per hour, but incorrect loading of the carousel may affect this.

Paperless picking can be achieved with the necessary level of computer control, by using light and LED displays showing both the location and the quantity to be picked.

A 7 m high system has been chosen because this maximises the economies of this type of system.

Table 3 - Costs of vertical carousel systems

2.2 Horizontal carousels

A horizontal carousel is a series of carriers which rotate horizontally to bring the required carrier to the picking station. Each individual carrier can hold several locations, with the shelves being adjustable in order to provide the necessary number and size of locations.

As all picking is carried out at the picking station this area can be well lit, whilst the rest of the warehouse needs only minimal lighting, for maintenance etc. The pickers walking time will also be reduced dramatically, therefore the picking rate can be high. Rates of 200 picks per hour have been claimed for simple picking. A more realistic level of 100 picks per hour has been used in this study to allow for difficulties in picking, maybe counting small items etc. as, in these situations 40 - 120 picks per hour is typical.

Table 4 - Costs for horizontal carousel systems

2.3 Mini-load AS/RS

A mini-load system consists of aisles of racking containing standardised trays at each location. The trays are often sub-divided to hold several smaller containers, in this case each tray holds 6 tote boxes. The trays are stored and retrieved automatically by a miniature stacker crane.

Mini-load systems can utilise considerable height in a building. For this example a height of just over 8 m has been used. Loading of the system is important, and a correctly loaded mini-load can achieve picking rates of 100 picks per hour, and in certain instances more.

Computer control is essential with a mini-load system to achieve high picking rates, but also the computer can control the inventory, reducing or even removing totally the need for time consuming stock-takes.

The costs shown in our example represent the most basic machine of this type and only include minimal controls. Much higher prices may be experienced if they are integrated with extensive conveyorised routing systems.

Table 5 - Costs for mini-load systems

Table 6 - Comparison of costs regardless of shifts worked

COMMENTS ON THE RESULTS

It is clear that at throughput levels below 250 accessions per day the less sophisticated manual pick is the answer as all our examples proved unfeasible below this level.

The free-path order-picker was eliminated from the winners list when we disregarded the number of shifts which had to be worked but as the calculations show it may be a winner if it is essential to work a specific number of shifts because of other operational factors.

At low ratios of throughput to stockholding the goods to picker methods take precedence but when the ratio increases the fixed-path order-picker is supreme.

It is very interesting to note the great range of cost per unit of throughput from 3.79p per item on a high throughput to low stockholding ratio to a high of 71.97p per unit where throughput is low and stockholding high. This emphasises the need to minimise range wherever possible.

Table 7 - Comparison of costs for one shift operation

Table 8 - Comparison of costs for two shift operation

Table 9 - Comparison of costs for three shift operation

CONCLUSION

The overriding conclusion of this study is that it pays to consider very carefully the interactions which occur between stockholding, throughput, capital and operating cost before selecting any given system.

One feature which is only partially reflected is that of reliability. It is obviously difficult to quantify the effect on a total system of the equipment not being able to deliver because of breakdown. Typically what might be the result if a computer failure renders the system inoperable for a period of time. Such assessments can be made but only in the light of knowledge of the requirements of a specific system and for this reason we cannot include this aspect in our calculations.

Finally we think it would be dangerous to draw any hard and fast conclusions from our tables of results. However, it is important to consider the use of this technique for any specific application in order to establish an adequate solution.

Table 1 Costs of free-path order picker system

COST DATA FOR FREE PATH SYSTEMS 7M HIGH

RENTALIZED COST IN POUNDS STERLING (£)

THROUGHPUT	500	500	500	500	500	1000	1000	1000	1000	1000	1000	5000	5000
HOLDING	2500	5000	10000	20000	20000	2500	5000	10000	10000	20000	20000	10000	20000
NO. OF MACHINES	1	1	1	2	2	1	1	1	3	1	3	4	4
NO. OF SHIFTS	2	2	2	1	2	3	3	3	1	3	1	3	3
WORKERS PER SHIFT	1	1	1	2	2	1	1	1	3	1	3	4	4
BUILDING COSTS	2978	4354	7257	13066	13066	2978	4354	7257	10310	13066	16272	11913	17413
RACKING COSTS	8179	16299	32398	64396	64396	8179	16299	32398	32398	64396	64396	32398	64396
MACHINE COSTS	4669	4669	4669	7472	7472	5952	5952	5952	11207	5952	11207	23810	23810
LABOUR COSTS	23100	23100	23100	21000	21000	39375	39375	39375	31500	39375	31500	157500	157500
EQPT MAINT	2640	2640	2640	3520	3520	3300	3300	3300	5280	3300	5280	13200	13200
CONTROL COSTS	3396	3396	3396	6367	6367	3396	3396	3396	7640	3396	7640	10187	10187
TOTAL	44961	54457	73460	115820	111266	63180	72675	91678	98335	129485	136296	249007	286505
COST COMPARISON PER ITEM THRU'PUT (PENCE)	34.59	41.89	56.51	89.09	85.59	24.30	27.95	35.26	37.82	49.80	52.42	19.15	22.04

DCF FACTOR (%)   11
RENTALISED BUILDING COSTS 8M HIGH (£ PER SQ.M)   61.66
RACKING COST PER LOCATION (£)   2.35
RACKING LIFE IN YEARS   10
TOTE BOX COST PER UNIT (£)   10.87
TOTE BOX LIFE IN YEARS   5
MACHINE COST (£)   22000
MACHINE LIFE IN YEARS   - 1 SHIFT   10
                         - 2 SHIFT   7
                         - 3 SHIFT   5
EQUIPMENT MAINTENANCE   - 1 SHIFT   8
AS % OF CAPITAL COST   - 2 SHIFT   12
                         - 3 SHIFT   15
LABOUR COST IN (£)   - 1 SHIFT   10500
                         - 2 SHIFT   11550
                         - 3 SHIFT   13125
CONTROL COSTS (£)   - 1 MACHINE   16000
                           2   "   30000
                           3   "   36000
                           4   "   48000
CONTROL EQUIPMENT LIFE (YEARS)   7
NO. OF DAYS PER YEAR   260

Table 2 Costs of fixed-path order picker systems

COST DATA FOR FIXED-PATH SYSTEMS 12M HIGH
RENTALIZED COST IN POUNDS STERLING (£)

THROUGHPUT	500	500	500	1000	1000	1000	1000	1000	5000	5000	5000	5000
HOLDING	2500	5000	10000	20000	2500	5000	10000	20000	10000	10000	20000	20000
NO. OF MACHINES	1	1	1	1	1	1	1	2	2	3	2	3
NO. OF SHIFTS	1	1	1	1	2	2	2	1	3	2	3	2
WORKERS PER SHIFT	1	1	1	1	1	2	2	2	2	3	2	3
BUILDING COSTS	2210	3364	5673	10290	2210	3364	5673	10290	11346	6729	7620	11346	12072
RACKING COSTS	8179	16299	32398	64396	8179	16299	32398	64396	64396	32398	32398	64396	64396
MACHINE COSTS	4019	4019	4019	4019	4450	4450	4450	4450	8037	10868	13350	10868	13350
LABOUR COSTS	10500	10500	10500	10500	23100	23100	23100	23100	21000	78750	69300	78750	69300
EQPT MAINT	960	960	960	960	1600	1600	1600	1920	1920	4480	4800	4480	4800
CONTROL COSTS	3396	3396	3396	3396	3396	3396	3396	3396	6367	6367	7640	6367	7640
TOTAL	29263	38537	56945	93560	42935	52209	70616	73479	107232	113066	139591	135107	176206	171558
COST COMPARISON PER ITEM THRU'PUT (PENCE)	22.51	29.64	43.80	71.97	16.51	20.08	27.16	28.26	41.24	43.49	10.74	10.39	13.55	13.20

```
DCF FACTOR (%)                                        11
RENTALISED BUILDING COSTS 13M HIGH (£ PER SQ.M)    78.15
RACKING COST PER LOCATION (£)                       2.35
RACKING LIFE IN YEARS                                 10
TOTE BOX COST PER UNIT (£)                         10.87
TOTE BOX LIFE IN YEARS                                 5
MACHINE COST (£)                                   32000
MACHINE LIFE IN YEARS    - 1 SHIFT                    20
                         - 2 SHIFT                    15
                         - 3 SHIFT                    10
EQUIPMENT MAINTENANCE    - 1 SHIFT                     3
AS % OF CAPITAL COST     - 2 SHIFT                     5
                         - 3 SHIFT                     7
LABOUR COST IN (£)       - 1 SHIFT                 10500
                         - 2 SHIFT                 11550
                         - 3 SHIFT                 13125
CONTROL COSTS (£)        - 1 MACHINE               16000
                         - 2 "                     30000
                         - 3 "                     36000
CONTROL EQUIPMENT LIFE (YEARS)                         7
NO. OF DAYS PER YEAR                                 260
```

Table 3 Costs of vertical carousel systems

COST DATA FOR VERTICAL CAROUSEL 7M HIGH
RENTALIZED COST IN POUNDS STERLING (£)

THROUGHPUT	500	1000	5000	5000	5000	5000	5000	5000	5000
HOLDING	1000	1000	1000	2500	2500	5000	5000	10000	20000
NO. OF MACHINES	2	2	2	5	5	10	10	20	40
NO. OF SHIFTS	1	1	3	2	3	1	2	1	1
WORKERS PER SHIFT	1	1	1	3	2	10	5	10	20
BUILDING COSTS	1308	1308	1308	2574	2574	4568	4568	9136	18271
RACKING COSTS	2940	2940	2940	7351	7351	14702	14702	29405	58810
MACHINE COSTS	2788	2788	4255	8511	10637	13940	17022	27879	55759
LABOUR COSTS	10500	10500	39375	69300	78750	105000	115500	105000	210000
EQPT MAINT	601	601	1403	2506	3508	3007	5012	6014	12029
CONTROL COSTS				INCLUDED IN MACHINE COSTS					
TOTAL	18138	18138	49282	90242	102820	141217	156804	177434	354868
COST COMPARISON PER ITEM THRU'PUT (PENCE)	13.95	6.98	3.79	6.94	7.91	10.86	12.06	13.65	27.30

```
DCF FACTOR (2)                                  11
RENTALISED BUILDING COSTS 8M HIGH (£ PER SQ.M)  61.66
TOTE BOX COST PER UNIT (£)                      10.87
TOTE BOX LIFE IN YEARS                          5
MACHINE COST (£)                                10024
MACHINE LIFE IN YEARS - 1 SHIFT                 15
                        2 SHIFT                 10
                        3 SHIFT                 7
EQUIPMENT MAINTENANCE  - 1 SHIFT                3
AS % OF CAPITAL COST     2 SHIFT                5
                         3 SHIFT                7
LABOUR COST IN (£)     - 1 SHIFT                10500
                         2 SHIFT                11550
                         3 SHIFT                13125
NO. OF DAYS PER YEAR                            260
```

Table 4 Costs for horizontal carousel systems

COST DATA FOR HORIZONTAL CAROUSEL 4M HIGH
RENTALISED COSTS IN POUNDS STERLING (£)

THROUGHPUT	500	500	1000	1000	1000	1000	1000	5000	5000	5000	5000	5000
HOLDING	1000	2500	1000	1000	2500	2500	5000	5000	5000	10000	10000	20000
NO. OF MACHINES	1	2	1	1	2	2	4	4	4	8	8	16
NO. OF SHIFTS	1	1	1	1	1	2	1	2	3	1	2	1
WORKERS PER SHIFT	1	1	2	1	2	1	2	4	2	8	4	8
BUILDING COSTS	1400	2801	1400	1400	2801	2801	5607	5607	5607	10786	10786	21588
RACKING COSTS	2940	7351	2940	2940	7351	7351	14702	14702	14702	29405	29405	58810
MACHINE COSTS	2086	4172	2086	2547	4172	5094	8344	10188	12733	16688	20377	33375
LABOUR COSTS	10500	10500	21000	23100	21000	23100	21000	92400	78750	84000	92400	84000
EQPT MAINT	450	900	450	750	900	1500	1800	3000	4200	3600	6000	7200
CONTROL COSTS	2334	2334	2334	2334	2334	2334	2334	2334	2334	4669	4669	9338
TOTAL	19711	28058	30211	33073	38558	42181	53788	128233	118328	149147	163636	214311
COST COMPARISON PER ITEM THRU'PUT (PENCE)	15.16	21.58	11.62	12.72	14.83	16.22	20.69	9.86	9.10	11.47	12.59	16.49

```
DCF FACTOR %                                     11
RENTALISED BUILDING COSTS 5M HIGH (£ PER SQ.M)   56.47
TOTE BOX COST PER UNIT (£)                       10.87
TOTE BOX LIFE IN YEARS                           5
MACHINE COST (£)                                 15000
MACHINE LIFE IN YEARS - 1 SHIFT                  15
                        2 SHIFT                  10
                        3 SHIFT                  7
EQUIPMENT MAINTENANCE - 1 SHIFT                  3
AS % OF CAPITAL COST    2 SHIFT                  5
                        3 SHIFT                  7
LABOUR COSTS (£)      - 1 SHIFT                  10500
                        2 SHIFT                  11550
                        3 SHIFT                  13125
CONTROL COST (£)        1-4 MACHINES             11000
                        5-8    "                 22000
                        9-12   "                 33000
                        13-16  "                 44000
CONTROL EQUIPMENT LIFE (YEARS)                   10
NO. OF DAYS PER YEAR                             260
```

Table 5 Costs for mini-load systems

COST DATA FOR MINILOAD SYSTEMS   8M HIGH
RENTALIZED COST IN POUNDS STERLING (£)

THROUGHPUT	500	500	500	1000	1000	1000	1000	1000	1000	5000	5000	5000	5000	5000
HOLDING	2500	5000	10000	2500	5000	10000	10000	20000	20000	5000	10000	10000	20000	20000
NO. OF MACHINES	1	1	1	1	1	1	2	2	1	2	3	2	3	2
NO. OF SHIFTS	1	1	1	1	1	1	1	1	2	3	2	3	2	3
WORKERS PER SHIFT	1	1	1	1	1	2	2	2	1	2	3	2	3	2
BUILDING COSTS	2222	2960	4585	2222	2960	5921	4585	9172	7831	4444	7549	5921	10654	9172
RACKING COSTS	8179	16299	32398	8179	16299	32398	32398	64396	64396	16299	32398	32398	64396	64396
MACHINE COSTS	8344	8344	8344	8344	8344	16688	10188	16688	10188	25467	30565	25467	30565	25467
LABOUR COSTS	10500	10500	10500	10500	10500	21000	23100	21000	23100	78750	69300	78750	69300	78750
EQPT MAINT	3000	3000	3000	3000	3000	6000	4200	6000	4200	9600	12600	9600	12600	9600
CONTROL COSTS					INCLUDED IN MACHINE COSTS									
TOTAL	32245	41102	58826	32245	41102	82006	74471	117256	109716	134559	152412	152135	187515	187385
COST COMPARISON PER ITEM THRU'PUT (PENCE)	24.80	31.62	45.25	12.40	15.81	31.54	28.64	45.10	42.20	10.35	11.72	11.70	14.42	14.41

```
DCF FACTOR (%)                                    11
RENTALISED BUILDING COSTS 8M HIGH (£ PER SQ.M)    64.14
RACKING COSTS PER LOCATION (£)                    2.35
RACKING LIFE IN YEARS                             10
TOTE BOX COST PER UNIT (£)                        10.87
TOTE BOX LIFE IN YEARS                            5
MACHINE COST (£)                                  60000
MACHINE LIFE IN YEARS   - 1 SHIFT                 15
                          2 SHIFT                 10
                          3 SHIFT                 7
EQUIPMENT MAINTENANCE   - 1 SHIFT                 5
AS % OF CAPITAL COST      2 SHIFT                 7
                          3 SHIFT                 8
LABOUR COST IN (£)      - 1 SHIFT                 10500
                          2 SHIFT                 11550
                          3 SHIFT                 13125
NO. OF DAYS PER YEAR                              260
```

Table 6   Comparison of costs regardless of shifts worked

```
SUMMARY RESULTS
(OUTRIGHT WINNERS REGARDLESS OF SHIFTS)
```

***	COST PER ITEM THROUGHPUT (PENCE)		***
ITEM THROUGHPUT	HOLDING	MIN. COST SYSTEM	MIN. COST
500	1000	V.C	13.95
500	2500	F.I	22.51
500	5000	F.I	29.64
500	10000	F.I	43.80
500	20000	F.I	71.97
1000	1000	V.C	6.98
1000	2500	M.L	12.40
1000	5000	M.L	15.81
1000	10000	F.I	27.16
1000	20000	F.I	41.24
5000	1000	V.C	3.79
5000	2500	V.C	6.94
5000	5000	H.C	9.86
5000	10000	F.I	10.39
5000	20000	F.I	13.20

```
V.C : VERTICAL CAROUSELS
H.C : HORIZONTAL CAROUSELS
M.L : MINI-LOAD SYSTEM (AS/RS)
F.I : FIXED-PATH ORDER PICKERS
F.R : FREE-PATH ORDER PICKERS
```

Table 7   Comparison of costs for one shift operation

```
SUMMARY RESULTS
(ONE SHIFT OPERATION)
```

***	COST PER ITEM THROUGHPUT (PENCE)		***
ITEM THROUGHPUT	HOLDING	MIN. COST SYSTEM	MIN. COST
500	1000	V.C	13.95
500	2500	H.C	21.58
500	5000	F.I	29.64
500	10000	F.I	43.80
500	20000	F.I	71.97
1000	1000	V.C	6.98
1000	2500	M.L	12.40
1000	5000	M.L	15.81
1000	10000	F.I	28.26
1000	20000	F.I	43.49
5000	1000	*	*
5000	2500	*	*
5000	5000	V.C	10.86
5000	10000	H.C	11.47
5000	20000	H.C	16.49

```
* : EITHER IMPRACTICAL OR INEFFICIENT

V.C : VERTICAL CAROUSELS
H.C : HORIZONTAL CAROUSELS
M.L : MINI-LOAD SYSTEM (AS/RS)
F.I : FIXED-PATH ORDER PICKERS
F.R : FREE-PATH ORDER PICKERS
```

Table 8   Comparison of costs for two shift operation

SUMMARY RESULTS (TWO SHIFT OPERATION)			
***	COST PER ITEM THROUGHPUT (PENCE)		***
ITEM THROUGHPUT	HOLDING	MIN. COST SYSTEM	MIN. COST
500	1000	*	*
500	2500	F.R	34.59
500	5000	F.R	41.89
500	10000	F.R	56.51
500	20000	F.R	85.59
1000	1000	H.C	12.72
1000	2500	H.C	16.22
1000	5000	F.I	20.08
1000	10000	F.I	27.16
1000	20000	F.I	41.24
5000	1000	*	*
5000	2500	V.C	6.94
5000	5000	H.C	9.86
5000	10000	F.I	10.39
5000	20000	F.I	13.20

* : EITHER IMPRACTICAL OR INEFFICIENT

V.C : VERTICAL CAROUSELS
H.C : HORIZONTAL CAROUSELS
M.L : MINI-LOAD SYSTEM (AS/RS)
F.I : FIXED-PATH ORDER PICKERS
F.R : FREE-PATH ORDER PICKERS

Table 9   Comparison of costs for three shift operation

SUMMARY RESULTS (THREE SHIFT OPERATION)			
***	COST PER ITEM THROUGHPUT (PENCE)		***
ITEM THROUGHPUT	HOLDING	MIN. COST SYSTEM	MIN. COST
500	1000	*	*
500	2500	*	*
500	5000	*	*
500	10000	*	*
500	20000	*	*
1000	1000	*	*
1000	2500	F.R	24.30
1000	5000	F.R	27.95
1000	10000	F.R	35.26
1000	20000	F.R	49.80
5000	1000	V.C	3.79
5000	2500	V.C	7.91
5000	5000	H.C	9.10
5000	10000	F.I	10.39
5000	20000	F.I	13.55

* : EITHER IMPRACTICAL OR INEFFICIENT

V.C : VERTICAL CAROUSELS
H.C : HORIZONTAL CAROUSELS
M.L : MINI-LOAD SYSTEM (AS/RS)
F.I : FIXED-PATH ORDER PICKERS
F.R : FREE-PATH ORDER PICKERS

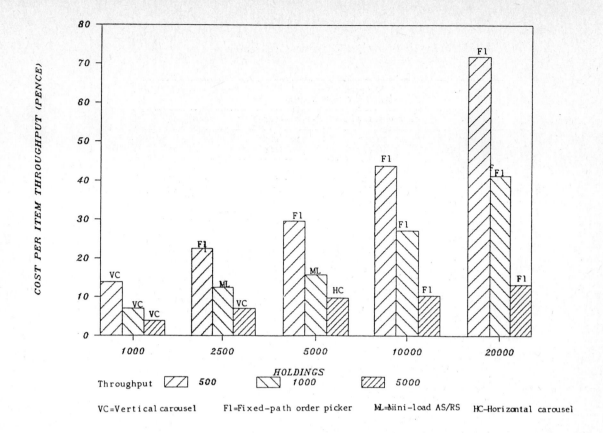

(a) Regardless of number of shifts

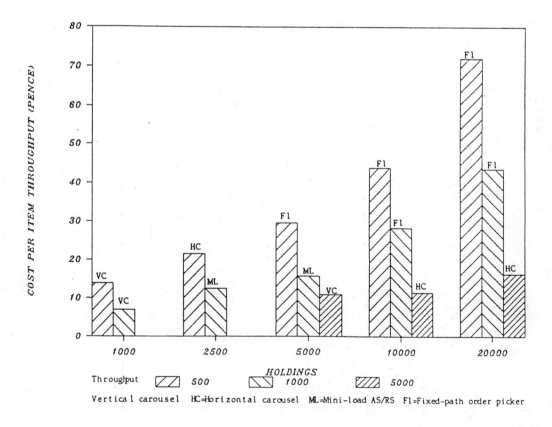

(b) On a one shift basis

Fig 1 Minimum cost systems

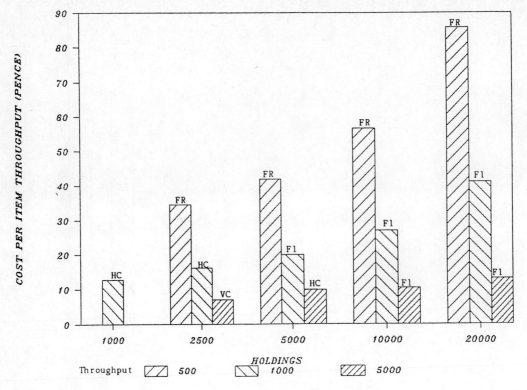

(c) On a two-shift basis

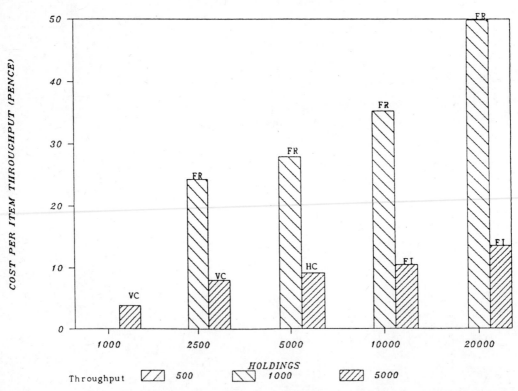

(d) On a three-shift basis

Fig 1 Continued

Appendix 1 Building cost relative to building height

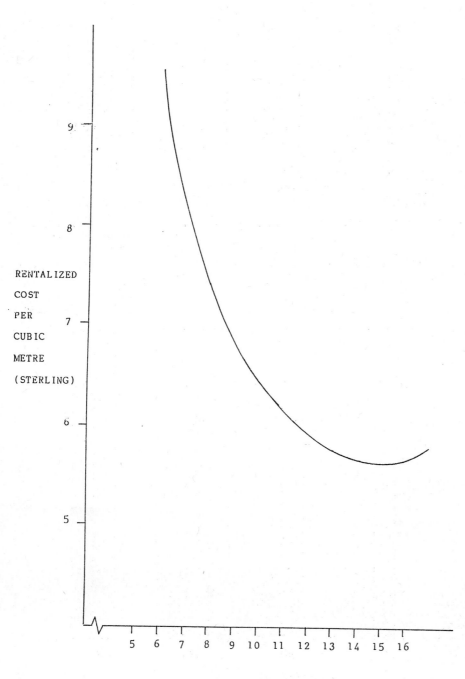

C88/88

The specification of advanced handling systems

R H HOLLIER, MSc, PhD, CEng, MIMechE, MIProdE, FIMH
Department of Management Sciences, University of Manchester Institute of Science and Technology, Manchester
G F SHIMMINGS
S M Consultants Limited, Windsor, Berkshire

SYNOPSIS

In essence an advanced handling system comprises mechanical handling equipment and computer hardware and software for control purposes. Having analysed its business strategy, a company may decide that it wishes to introduce an advanced handling system as one element of its overall manufacturing and/or distribution activities to remain competitive. During the project management cycle it will become necessary to produce a requirement specification for the system and later to write detailed functional specifications for all the sub-systems. The aim of this paper is to provide guidance on the preparation of requirement specifications for advanced handling systems.

1 INTRODUCTION

Over the past fifteen years the technology involved in handling and storing materials in unit loads has made considerable progress. During the late 1970's there was much talk of the 'systems approach', then came the emphasis on flexibility and more recently the theme of integration. An advanced handling system embodies these concepts and comprises mechanical equipment and associated computer hardware and software for control purposes. As a result of these developments, the process of specifying such systems, procuring them and installing them successfully has become much more complex. Yet the expertise available in most companies to carry out these tasks has not kept pace so that external advice usually has to be obtained. The aim of this paper is to give some overall guidance on the preparation of requirement specifications in the context of managing a project to implement an advanced handling system (AHS).

Little has been published on this subject, even in specialist books on materials handling.[1] In 1974, the trade association CAMA (now GAMBICA) produced a booklet[2] on the conduct of projects involving the combined use of mechanical and control equipment. There have also been some papers dealing with specifications for the design of new products or for purchasing, for example those by Pugh[3] and Macdonald.[4] Very recently the British Standards Institution has formed a new drafting committee to update PD 6112[5] and it has also produced a valuable standard on the specification of computer software[6] which is part of the problem addressed here. The present work came about as a result of a number of meetings between equipment users and suppliers organised by the British Materials Handling Board where it became apparent that both sides would welcome the availability of a guidance document on this subject, a view shared by the recently formed Automated Materials Handling Systems Association. The full document is shortly to be published by IFS Ltd.

2 STRATEGY AND OBJECTIVES

Before a project is started a company will need to give consideration to its overall business strategy which determines its response to the market and its competitors. This strategy will have many facets in terms of markets and products, manufacturing and logistical activities, financial and commercial aspects and so on. The strategy may well demand some improvement in the logistical performance of the company, for example, in shortening lead times to give better customer service or in decreasing stock levels to reduce unit costs. This, in turn, will provide objectives for an initial feasibility study to evaluate whether such an improvement could be obtained by the introduction of new systems of handling and storage using advanced technology. If the outcome of this feasibility study is positive, then a full project will follow with a set of well-defined objectives.

Therefore, a clear statement of objectives is the first positive step in the process of creating and communicating project specifications; it is the cornerstone of the requirement and the datum against which all other achievement must ultimately be measured.

3 PROJECT TEAM

Those in management who believe that technology provides one primary means of ensuring their future will already have established a personal, departmental or corporate interest in automation and the current 'state-of-the-art'. It will soon be apparent that no single person will have the expertise, experience and capability to design and implement an advanced handling system (AHS) to perfection and that a project team is a necessity to the successful pursuance and conclusion of every AHS project.

The project team will need a leader who can bring drive and enthusiasm to the mission, someone well versed in business strategy with intimate working experience of corporate operating requirements and practices. The leader must have integrity, the ability to motivate and co-ordinate people and the perception to bring in essential expertise as required. The person must plan and document events with firmness and discipline to reduce the incidence of delay and to achieve high standards of quality and performance within budget.

The selection of the project team itself is another key element in the process of creating and communicating project specifications. Firstly, it will be necessary to identify those skills which the project demands and which must be present in the team from the start. Then internal skills, their source and availability, must be assessed. By matching these two, any skill deficiencies can be identified and steps taken to obtain external assistance. Input of the very best advice, particularly during the early stages, is essential, as is the need to maintain adequate levels of technical competence throughout the project, bringing in specialist skills for short-term involvement. The team will need to be small in number and given adequate time to work on the project relieved, as far as possible, from other duties.

The project team will provide the organisation for the project through all its stages and ensure continuous assessment of progress. However, its most important and crucial task is to prepare the project specifications.

4 PROJECT STAGES

A typical AHS project can be divided into 20 stages (see Table 1). The form and content of each stage will vary according to the type of project, but it does provide a basic checklist against which progress can be monitored. In concentrating on the preparation of project specifications, the main emphasis of this paper will be on Stages 1 to 3, although some reference will be made to later stages, especially Stage 10.

TABLE 1: PROJECT STAGES

Stage 1 Development of the operational specification

2 Preliminary system designs, layouts and functional specifications

3 Testing and modification of preliminary designs

4 Budget costing

5 Capital approval

6 Preliminary project organisation

7 Selection of tenders

8 Evaluation of tenders

9 Placing of contracts

10 Finalisation of the functional specification and design detail

11 Contract management and co-ordination

12 Site management

13 Equipment proving at suppliers

14 Installation and maintenance training

15 Commissioning and preliminary operator training

16 Equipment and systems proving

17 Acceptance testing

18 Full instruction and training

19 Full operation

20 Development and corporate integration

From Table 1 it will be noted that a distinction is made between an operational specification (sometimes termed a requirement specification) and a functional specification. In the case of an AHS, the operational specification defines the detailed characteristics of the unit loads to be moved and stored, throughput rates and capacities, interfaces with other functions, overall control systems, constraints such as existing buildings, etc. On the other hand, the functional specification defines the detailed control systems which will be embodied in the computer software and its interfaces with human operators. Later, mention is also made of equipment specifications.

5 STAGE 1 - DEVELOPMENT OF THE OPERATIONAL SPECIFICATION

The initial five stages of a project are fundamental to its successful realisation, for they establish the foundation of the entire project and determine initial technical feasibility and economic viability. The development of the operational specification is the most important of all, for it is the precise definition of a total operational requirement, in fulfilment of the established project objectives. Of necessity the project team will spend a great deal of time and painstaking effort on the preparation of the operational specification.

a) Activity Profile

The primary and obvious need is to establish an accurate and meaningful profile of the activity to be automated, together with all other activities which are directly or indirectly associated with it.

b) Operational Data

The development of a basic movement diagram will rapidly enable the initial identification of required statistical and operational data. It is at this stage that the project team will require the patient co-operation and involvement of other corporate disciplines, some of whom may need to be co-opted as additional members.

Some of the disciplines most likely to be involved include:-

- Marketing   All product line data, historic and projected sales, new and discontinued lines, promotional activity, number and type of outlets, etc.

- Production   Information concerning existing and proposed plant and processes, machine cycle times, interfaces, plant layout, etc.

- Operations & Transport   All data and information concerning warehousing and distribution.

- Quality Control — Constraints and other handling restrictions associated with the production and transportation of product.
- Packaging — Physical dimensions, weights, pallet quantities, and information concerning projected or discontinued cartons.
- Purchasing — Any special provisions required for the storage and handling of bulk purchased materials.
- Health & Safety — Environmental and operator constraints.
- Data Processing — The extraction, provision and interpretation of essential data.
- Board of Directors — It is important that the project team be privy to corporate strategic planning, including any plans for expansion or contraction, in so far as they may affect the project.

The collection of project data is often a tedious and difficult task. The information particularly required is frequently the most difficult to extract, or so it seems!

The need for accurate data at this stage cannot be over-emphasised:

- garbage in - garbage out
- guesstimates are dangerous
- if it is not available, try the production or warehouse foreman; he may have it in his "little black book"
- be prepared to carry out a special study or analysis if necessary
- identify peak situations - by month, week, day and hour
- most peaks are market-orientated - be doubly alert to those that are operationally induced but are unnecessary
- thoroughly investigate the possibilities of operational "smoothing" of peaks
- beware of peak-on-peak situations, they may invalidate a project on economic grounds
- thoroughly examine the marketing plan and define its impact upon every aspect of the project over the next five years at least
- be prepared to discover and eradicate inefficiencies and illogical practices that may well have built up over a period of time (an added corporate benefit)
- take pride in providing accurate production, movement and storage profiles.

c) Preliminary operational specification

This is the most important document of the project and the ultimate key to the successful implementation of an AHS. It should be prepared on a large sheet with a conventional symbolised movement diagram to one side and the legend on the opposite side should contain all of the operational data pertinent to each particular section. This document is a statement of fact - it states what has to happen, not how. It should be developed and carefully thought out by the project team, discussed and ratified by all interested parties. It must include all operational requirements and provide the important medium for involvement and agreement with all disciplines. Above all, the operational specification is a statement of requirements, not the definition of a method of solution or an outline system design.

6 STAGE 2 - PRELIMINARY SYSTEM DESIGNS, LAYOUTS AND FUNCTIONAL SPECIFICATIONS

The project team should now turn its attention to the system design, layout and functional specification. Apart from dimensions, performance details and guide pricings, suppliers are best consulted at a later stage, when there is a greater awareness of what is required and what is being offered. Other installations are unlikely to have more than a passing relevance since their requirements are bound to be different.

a) Layouts and Schematics

The skills necessary to transform the operational specification developed in Stage 1 into system options and functional specifications are not easily acquired. It is at this stage that the best available expertise from each involved discipline, together with co-opted and/or imported specialists, will need to be employed. The objective is to produce a number of schematic equipment options all of which fulfil the operational specification.

The following points should be borne in mind:

- Layouts and schematics will embody the criteria established during Stage 1
- Particular attention must be paid to the performance requirements of all production plant and materials handling equipment
- Carefully plan the siting of equipment and the sizing and location of all internal areas to maintain the most efficient flow of materials
- Incorporate adequate storage areas, particularly for 'Work-in-Process' and 'Finished Goods'
- Some of the essential areas and activities which may be neglected or even overlooked are:-

Handling and storage of empty stillages and pallets

Handling and storage of packaging and waste materials

Jig and tool handling and storage

Component handling and storage in assembly areas

Maintenance and charging areas for handling equipment

Provision of adequate access for plant maintenance

Inability to expand in a key production area

Inadequate marshalling areas

b) Building Options

It is only when these layouts and schematics are completed and all requirements have been determined, that final consideration should be given to the building. Always remember that the best and most economic option may involve demolishing existing buildings or even moving the operation to a new site. Some initial consideration will already have been given to the building requirement during the schematic design stage; however, this should not be allowed to impose constraints on the design before each option has been evaluated. In a manner peculiar to each individual situation, the options which need to be evaluated will emerge. Each option will be influenced by a number of varying conditions, some of which are:-

- Geographical location
- Availability and cost of land
- Constraints on existing buildings
- Grant aid
- Multi or single storey operation
- High or low bay potential
- Specialised building requirements
- Building costs
- Equipment costs

Reputable building contractors and equipment suppliers are usually willing to assist at this stage by providing reasonably accurate estimates so that each available option may be totally evaluated. It is desirable that, at this particular point in their deliberations, the project team should have positively identified the preferred option in terms of location, building outline and system layout.

c) Preliminary Functional Specification

Once more, it should be remembered that this is a specification of requirement rather than a definition of the method of solution. Based upon the preliminary operational specification, system design and building layout, the preliminary functional specification must define in broad terms, the format of the control system.

The project team will need adequate knowledge and expertise in order to prepare this specification. If this is not available then it must be co-opted internally or brought in from external sources.

(i) Main Requirements

- A control philosophy will need to be discussed and developed in conformity with the preliminary operational specification.
- All system functions must be clearly defined. This will include system logic and all other control functions such as documentation, identification and coding, specialised sensing, diagnostics, etc.
- Inputs and outputs should be identified
- All operator interactions must be specified including associated hardware, eg keyboards, VDU's, printers, etc.
- Interfaces with other equipment or systems must be defined
- Consideration must also be given to any 'back-up' equipment which may be required
- The preliminary functional specification should be one which can be met in the most economic manner utilising proven equipment and knowhow.

(ii) Computer Hardware and Software

In support of the main requirements listed above, a number of factors need to be taken into account when specifying the computing element of the proposed system. Some are of relevance to the preliminary stages; all are of importance prior to any contract finalisation.

The choice of computer hardware is often made to appear too important. This is not so; there are many suppliers who have suitable equipment. However, purely commercial multi-user computers, even if they appear to have interactive on-line facilities, should be avoided; only use computers and equipment that is widely employed in on-line control applications. Suitable equipment already installed may influence, but should not dominate, the choice of new hardware. The type of computer equipment chosen should not determine the supplier of software.

In most advanced industrial control systems, there is a hierarchy of computers dealing with different control levels. Thus at the top level there will be a corporate computer, below which there will be a number of system control computers including that for an AHS, while at the bottom level individual pieces of equipment will have their own Programmable Logic Controllers (PLCs) or microcomputers for local control purposes. The functional specification will usually need to cover the bottom two levels of the hierarchy and the interfaces to other system computers and the corporate computer. If these interfaces allow proper communication, then potentially the major benefit of integrated operation of the entire business from planning to execution can be derived from computer control.

The key element in the whole system will be the software. Because software engineering is a relatively new discipline and not well understood, it is often given inadequate attention even at the functional specification stage. It is only later when changes in the control system are required, that it is realised that there exist major constraints imposed by the design of the software. The provision of quality software, which is flexible, adaptable and easily maintained, is such a specialised business, that it is usually advantageous to use a reputable software house or a group supplying a major part of the mechanical equipment. Even then considerable effort will need to be given by the project team to defining in detail the functional specification so that there is no misunderstanding with the supplier, who will probably attempt to use his standard software products if possible.

(iii) Sizing the Computer

Accurate sizing of the computer is vital to every successful project. Meticulous identification of the various functions to be performed is, of course, the most important aspect, for these will clearly define the scope of the system and determine how much of the operation will come under the control of the computer system.

It is equally important to determine what speed of response is necessary for different functions. Selected functions will require immediate response in order to capture the throughput potential of the mechanical equipment. However, with careful planning, most operational functions can be anticipated to allow the computer to prepare the next operation instruction signal before the particular piece of equipment calls for it.

Fundamental to the design of the software is the analysis of data flow. For example, certain linkages will only need small quantities of rarely changed data, while others will be concerned with passing frequently changing status information to the system computer. Another factor is the effect of system failure which will dictate modular software design to ease recovery. These considerations will determine the type of network required and its associated systems software.

A decision will need to be taken as to whether the computer system is to be duplicated, or whether the finite reliability of a single computer is acceptable. If a dual system is selected, then the question arises of whether it should run as a 'hot standby', ready to take over immediately if its counter-part fails (possibly with automatic switching), or whether a delay of 15 to 20 minutes is imposed. The latter is more likely to be acceptable in the mechanical handling environment.

Once the basic functions have been defined, consideration must be given to the quantity of peripheral equipment which is required, eg. VDU terminals, printers, bar code readers, etc. and also to the layout of on-line mechanical equipment and the number of associated PLCs and local controllers. These many considerations combine to determine the number of serial connection 'ports' on the system computer.

The number and size of files required will need to be determined, for this is an important element in sizing the computer. Apart from the obvious control files, such as the stock file and production schedule, there are frequently many other subsidiary files which may be required for management purposes, such as transaction recording and management statistics. Enquiry and reporting functions from the computer controlling the equipment should be restricted to the bare essentials and should not be given priority over control functions.

d) Financial Element

The completion of the preliminary operational specification, system design, building layout and preliminary functional specification will enable discussion of the project with two or three suppliers to begin and provide them with sufficient information for them to be able to offer budget costs to an accuracy of $\pm$ 10%, excluding risk.

An important requirement of Stage 2 is that the preliminary design concept(s) should not only anticipate an acceptable return on investment, but also be affordable within the prevailing corporate financial climate. An overall awareness of costs on the basis of a continuous assessment is an essential responsibility of the project team. It would seem pointless to pursue non-viable designs, and yet, it is surprising how many projects are advanced with no hope of Board approval.

e) Other Considerations

There are many other considerations which may impact the total project and therefore require attention. Some of these include:-

Civil and building regulations and requirements

Consents and permissions

Local by-laws and regulations

Environmental requirements

Fire regulations and F.O.A. recommendations

Floor loadings and other structural stresses

Health & Safety/Factory Acts

Industrial relations

Scheduling

Site access and headroom, etc.

Installation (corporate requirements)

Power and other service requirements

f) Risk Factor

The extent of the risk involved and the size of the contingency for risk which is to be added to the capital cost will depend upon the extent and complexity of the project and the knowledge and expertise of the supplier and his contractors. The application of an overall percentage to cover risk should be avoided. It is more prudent to analyse and identify specific areas of risk and then to apply an individual figure to each item. This practice will assist budget control during project implementation and signal elements of the contract to receive special attention during these stages.

7 STAGE 3 - TESTING AND MODIFICATION OF PRELIMINARY DESIGNS

a) Departmental Presentations

This most important and frequently neglected stage in the progressive development of a project affords the project team the formal opportunity to review, with management and all of the operational departments, the proposed scheme.

The objectives of this stage are to ensure fulfilment of all operational requirements and to ensure that all interfaces, mechanical, electrical, manual and computer, have been identified and defined. Furthermore, it gives an opportunity to eliminate any misconception and doubts concerning the operation and function of the system and to engender agreement, support and commitment within the corporate organisation.

By actively pursuing this policy of discussion with all interested parties, the project team will preempt most of the late changes and misunderstandings which are so frequently the cause of unforeseen delays, additional costs and post-contract conflicts with suppliers. The time and effort expended on this stage will be worthwhile, for it will improve the quality and tenor of the project, maintain budgets and help to achieve objectives and schedules.

Although this departmental approach will ensure that each corporate discipline will review the project, the two main areas for particular attention remain the mechanical and computer elements.

b) Mechanical Element

The inclusion within the system design of new and untried equipment, or standard equipment utilised in an entirely new application or environment, will demand special attention and evaluation. A potential supplier will need to unequivocally demonstrate efficient and sustained performance, reliability, and acceptable levels of maintenance and running costs.

Particularly careful consideration should be given to any new equipment upon which an entire operation is totally dependent. Whereas industry needs to avail itself of new technology and techniques, an operating company can seldom afford the potential hassle and cost of becoming the "field development installation" of a supplier.

Under certain conditions, investment in a prototype is frequently beneficial and may not be expensive, particularly when viewed against potential savings. It is unusual for a complete machine or piece of equipment to need to be constructed - rather for one particular function or mechanism to be built and tested.

This stage also offers the opportunity to carry out trials and tests of manual interfaces utilising simple basic equipment in association with trained and experienced operatives whose co-operation and involvement is usually found to be enthusiastic and constructive. The objective is to ensure that the proposed arrangements will work. Some typical manual interface examples include machine load/unload, assembly operations, selection, packing, palletisation, vehicle loading and computer/control interaction.

There are many operational simulations, whether manual, mechanical or computer based, which are essential and always of value, provided the inputs are realistically constructed and any sensitivity to variation in key parameters is tested. Simulations should certainly be provided where any form of operational sequencing and integrated movement is involved or where any form of automated or semi-automated order selection is used.

c) Computer Element

The entire computer/control package will have been subject to scrutiny by all corporate departments. The principal requirements of this stage are to identify clearly the scope and functions that are necessary, stating these in terms of understandable and simple requirements. It must also be possible to visualise the operation, providing sufficient control and operational information to enable the computer element to be thoroughly understood by all concerned. It must be appreciated that special software will usually be required if the system incorporates mechanical novelty or complexity. The system may be elaborated later, but may fail if it becomes too complex for the initial installation when everything is new.

8 STAGES 4 to 9 - BUDGET, APPROVAL, TENDERS AND CONTRACTS

The finalised operational and functional specifications will be used in the budget costing and capital approval Stages 4 and 5. They will then be used again for tendering purposes in Stages 6, 7 and 8. As a result of the evaluation of tenders in Stage 8, it will be necessary to finalise the equipment specification prior to the preparation of contracts in Stage 9. A vital part of the contractual agreement will be the establishment of acceptance criteria for the overall performance of the AHS once it is in full operation.

9 STAGE 10 - FINALISATION OF THE FUNCTIONAL SPECIFICATION AND DESIGN DETAIL

The better the tender documentation, incorporating the operational and functional specifications, the less there will be to finalise prior to manufacture. The project has been approved and sanctioned and supplier(s) appointed. There must now follow a period of finalisation and re-checking in order to ensure that every requirement has been identified, understood, documented and notified to all parties. If necessary, revisions must be agreed. Both purchaser and supplier(s) have equal responsibilities in so far as each must thoroughly understand what is being supplied, when and how, and for what price. This stage must not be neglected for it is the last opportunity to discover those elusive factors which have a habit of impacting on contract price, schedule and performance.

10 ACKNOWLEDGEMENTS

The authors gratefully acknowledge the work of Mr Wes Spooner of SM Consultants Ltd. who contributed to the document referred to in the Introduction on which this paper is based.

11 REFERENCES

1. Kulwiec, R A (Ed). Materials Handling Handbook. 2nd Ed, Wiley. 1985.

2. Control and Automation Manufacturers Association. A Guide to the Procurement of Complex Electronic Control and Supervisory Systems. CAMA, 1974.

3. Pugh, S. Preparation Material for Design Teaching: Specification Phase. Sharing Experience in Engineering Design, 1986.

4. Macdonald, R M. Drawing up the Purchasing Specification. Proc.Instn.Mech.Engrs.,Vol 199, pp41-45, 1985.

5. PD 6112. Guide to the Preparation of Specifications. BSI, 1967.

6. BS 6719. Guide to Specifying User Requirements for a Computer-Based System. BSI, 1986.

C89/88

A case study of the application of computer simulation to automated storage and handling systems

A J KEITH, BA, CEng, MIMechE, MIProdE, MBIM and **K D PORTER**, BSc, MA (Consultant)
Touche Ross Management Consultants, Guildford, Surrey

SYNOPSIS

A discussion of a simulation exercise for a new distribution centre. The simulation task was relatively complex and combined automated and manual systems. The simulation was used as a system design tool, and for modelling alternative business expansion strategies.

BACKGROUND

The computer simulation exercise was carried out as part of the design phase of the new distribution centre for Baxter UK Limited, a medical supplies company. Baxter was moving from regional distribution warehouses to a centralised facility. The company was especially suited to automating this facility because of the strict quality control requirements, and prior implementation of an extensive physical inventory control system.

The simulation model was developed during the final design phase of the project. It was necessary to have reached a stage at which further design work was seen to be 'fine tuning' work, and extensive data has been analysed to generate the daily work load on the model. Due to overall project timescales, experimentation was carried out during the equipment tender phase. This enabled results to be used in developing the detailed functional specification with the chosen contractor.

PHYSICAL SYSTEM

In order to understand the simulation task, it is necessary to briefly introduce the physical system, a block diagram of which is shown in Fig. 1. For simplicity, we only discuss those areas included in the simulation model.

The main functional areas of the Centre are:-

- Intake
- Storage cell
- Forward picking
- Load accumulation

The Intake Area receives all goods into the Centre. The goods proceed to the automated Storage Cell by conveyor system. Goods are received on a combination of random and scheduled deliveries.

The Storage Cell holds 96% of stock in the Centre.

The Centre is capable of handling emergency medical orders with immediate response. A design requirement was developed that a representative stock of every product must be held in a manually accesible location. This stock is held in Forward Picking.

Orders are processed by consolidating a shift's requirements by product code and then issuing the consolidated quantities to the Load Accumulation area where they are manually sorted to order pallets, by operators driving electric trucks.

Generally, full pallet requirements are issued automatically from the Storage Cell and part pallet requirements are picked and issued from Forward Picking.

All manual operations, as well as the automated equipment are under the control of the warehouse control computer system, manual operations being directed by radio terminals on the trucks.

Over the course of a shift, all orders are built up, product line by product line until complete.

SIMULATION METHOD

The simulation model was developed using OPTIK, a FORTRAN Based Simulation Tool with interactive facilities. The approach to the simulation was in four phases:-

- Specification
- Model building
- Experimentation
- Training

Significant benefits were derived at all of these stages, and it will be seen that the benefits of the simulation exercise were greater than the formal results of the modelling.

MODEL SPECIFICATION

A formal specification of the model was drawn up. The subjects covered by this specification were:-

- The extent of the model
- Control rules for each element of the model
- Variable parameters
- Reports to be generated

The model was chosen to represent the manually operated functions in the Centre in detail on the graphics display, whereas the automated pallet cranes were displayed in summary. The cranes carried out storage and retrieval cycles calculated on the basis of randomly generated locations, with various activities displayed in first form.

The functions covered were restricted to those areas described above. Several other less critical areas were omitted from the model.

The reason for this choice of extent of the model was to strike a balance between model development time, model speed and a focus on the areas of design interest against model completeness.

A graphics screen format was derived that allowed the majority of functions to be displayed, (see Fig. 2).

It was decided to run the model on the basis of randomly generated daily requirements, based on a derived profile of the business. This allowed the model to be run extensively for consecutive days, at different activity levels.

This method was chosen in preference to running the model always on the same 'typical' day, which could mask potential problems

The control rules for each element were defined. It was necessary to establish priorities for all eventualities and the method of operation for all equipment types. This process developed the scheme to a much greater level of detail at a relatively early stage in the project.

For example:

Pallets feeding the Load Accumulation Area arrive from the Storage Cell and Forward Picking. Both are serviced by the same Sortation operators. Which area should take priority? Does this priority change as buffer capacities are filled up?

Different equipment types could be used for picking and stacking operations in Forward Picking, or dual purpose trucks could be used. Any combination of these was required to be modelled. The work allocation and equipment interaction rules were defined.

The approach to replenishment of Forward Picking was also changed at this stage. It was decided that stock would be replenished via Sortation, eg.:

- Forward Picking Stock   0.25 pallets
- Total Consolidated Shift
  Requirement: 3.5 pallets

Action:

- 0.25 pallets from Forward Picking
- 4.0 pallets from Storage Cell to Sortation
- 0.75 pallets from Sortation to Forward Picking

This method reduced handling movements and integrated replenishment operations.

MODEL BUILDING

After agreement to the model specification by all parties involved in the project, model building commenced.

During early runs of the model, it was discovered that serious congestion could occur at the Forward Picking pick-up and deposit buffer locations (P & D's)

It was necessary to schedule the activities in such a way that sortation operators were available to clear the P & D's, and reduce the priority of pallets waiting on the output conveyor.

An additional report was added, which showed the level of utilisation of P & D's through the shift.

EXPERIMENTATION

The model was run for different activity levels, representing the start-up conditions and projected activity after five years.

Experimentation encompassed:

- Different equipment levels
- Alternative equipment combinations
- Equipment breakdown
- Various priority changes
- Product range changes

It was possible to run a complete eight hour shift simulation in eight minutes. It was therefore very practical to carry out experimentation work.

Sample results tables are attached

This phase enabled equipment types and quantities to be finalised, controlled rules to be agreed, and operating procedures for future growth to be developed.

TRAINING

The development of animated graphic simulation allows the model to be used to demonstrate the operation of the system. The model was used to present the system to all levels of Baxter staff, from truck operators to board directors.

The derived benefits of confidence and understanding of the project across a wide range of personnel are difficult to quantify, but were of at least equal value to all the earlier stages.

CONCLUSIONS

It is clear that the project benefited in many ways from the appropriate use of computer simulation.

There were, however, key features in the approach to the project that ensured that the benefit was obtained.

1. The system and operating procedures were largely defined before simulation commenced.

2. A thorough specification was written and agreed before any programming commenced. This procedure advanced the design to a detailed level before any modelling was carried out.

3. Variable parameters were mostly identified in advance of modelling. This enabled quick changes to the model during experimentation, by appropriate structuring of the model.

4. The extent of the model was restricted to critical areas. It was not an objective to produce a comprehensive model of the Distribution Centre.

5. The end-user and the modeller worked closely together. This avoided divergence from the original objectives.

6. The use of an animated display was exploited as a communication tool.

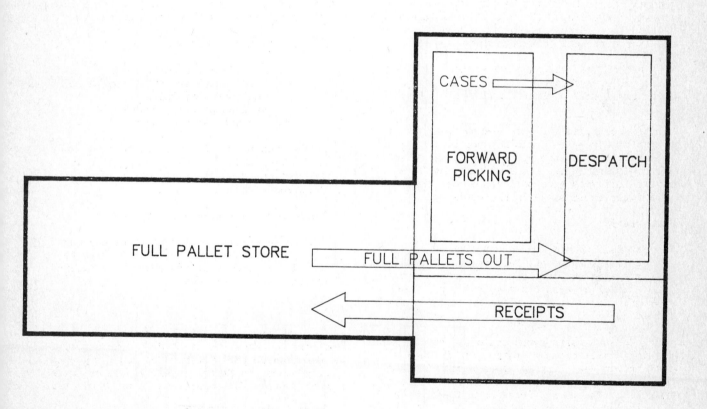

Fig 1 The physical system

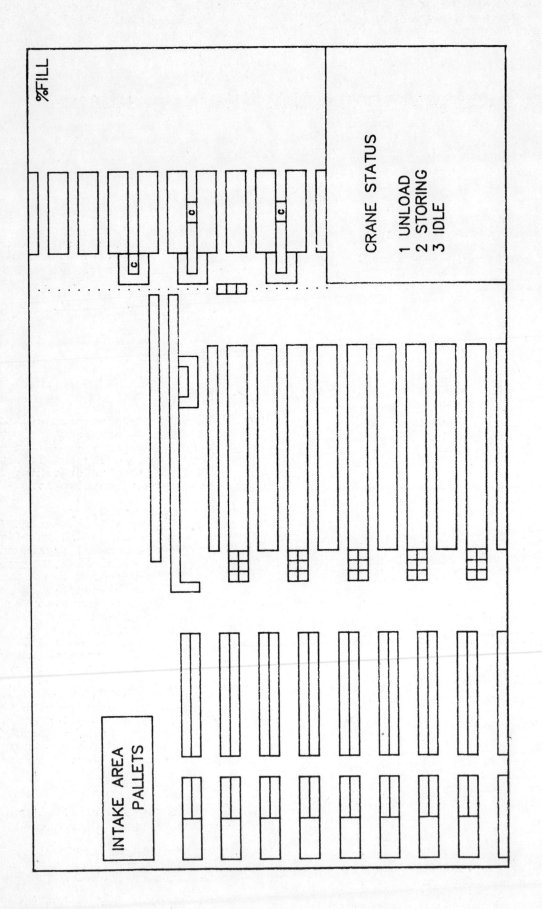

Fig 2 Functions of the model

# C90/88

Project management — the vital ingredient

S J TOMLINSON, BTech, MBIM, MIMM
Topstore Warehouse Systems, Slough, Buckinghamshire

SYNOPSIS Good project management is one of the vital ingredients of a successful project and this is especially true in the case of automated materials handling systems. More problems with such systems can be attributed to poor project management than almost any other factor. This paper will seek to discuss the vital importance of highly professional project management in the supply of large complex systems. Reference will be made throughout to existing installations to illustrate the points made.

1. INTRODUCTION

Projects can be described as:
'Any human activity which achieves a stated objective in a stated time within a stated budget'.

There are three major categories of projects:-

Manufacturing Projects -
The production of a piece of equipment involving design, testing and actual manufacture. The bulk of the work is carried out on the manufacturers premises.

Installation Projects -
The aim is to establish new buildings or operating plant at a remote site.

Management Projects -
Typically involve setting up new systems within an existing organisation.

Automated materials handling projects will normally include all three of these activities and for this reason tend to be very demanding and complex. Furthermore, projects will often run for years rather than months thus making planning and control even more complicated.

This paper seeks to present and discuss Project Management as it is carried out by TOPSTORE Warehouse Systems. To illustrate some of these points reference will be made to a number of projects carried by TOPSTORE.

2. BACKGROUND

On the basis of more than 20 years experience in a variety of automated materials handling projects considerable expertise in project management has been accrued. Furthermore this experience not only covers completely new factory installations but also refurbishing and modernising existing installations.

For a company which both manufactures and installs the majority of the equipment supplied and is often responsible for the general management of a project it is vital that the same rigorous project management techniques are applied to in-house manufacturing as during the more visible installation phase. Time delays and additional expenses incurred early on have a 'knock-on' effect and can seriously affect the outcome of a project which may be worth many millions of pounds.

For this reason Project Management is clearly one of the most important aspects of a major installation and it is on the strength of Project Management abilities that a considerable proportion of a companies success lies.

3. PROJECT MANAGEMENT

A project, as defined earlier, leads to the definition of project management:-

'The act of controlling any activity which achieves a stated objective in a stated time within a stated budget'

In the TOPSTORE organisation projects are broadly split into two phases, sales (Acquisition) and implementation. As they tend to require rather different approaches each phase maybe handled by separate personnel. This allows the appropriate engineers to concentrate on their chosen aspect of management.

The two functions, sales and implementation, have an important overlapping role as shown in the project life cycle diagram attached.

The remainder of this paper concentrates on the implementation phase.

4. THE PROJECT MANAGER

It is essential that one person is appointed as project manager and has total responsibility for the project.

A Project Manager should have :-

- Organisational talent
- Planning ability
- Leadership qualities
- Decision making skills
- Subject knowledge
- Self assurance
- Communication skills.

As a specialist he must be :-

- Systematic and rational
- Quick to understand
- Able to learn new technology
- Able to apply his specialised knowledge in a practical way
- Adaptable to each customer's style of operation.

Having established the ideal personal profile for a project leader the tasks he undertakes during a project can be briefly summarised as follows :-

Project analysis

- Determine customer requirements

Project aim

- Establish detailed specification

Project Organisation

- With customer
- With 3rd party suppliers

Project Costs

- Overall project costs
- Payments to 3rd party suppliers
- Payments from customer
- Bank guarantees
- Penalty conditions
- Monitoring changes from agreed specification

Project planning

- Overall project plan (what, when, how)
- Detailed plan
- Financial plan
- Budget
- Effects of National and Company regulations
- Meeting plan

- Co-ordination with customer and suppliers

Control of project

- Liaison with customer and third parties
- Monitor performance in accordance with the order and detailed specification
- Documentation of variations
- Documentation of progress
- Safety regulation enforcement
- Conformance with specification

Close of project

- Documentation
- Closing reports
- Service guarantees
- Closing accounts
- Training

5. DETAIL PROJECT SPECIFICATION

Any project must start with a specification as it is on this that the whole of it will be based.

Producing a project specification always involves close liaison with the customer.

The specification clearly establishes all the objectives which have to be reached by the conclusion of the project and in particular the following are covered in great detail :-

- Functional Description
- System Performance
- Timescales
- Scope of supply
- Technical standards
- Interfaces
- System testing
- System hand over
- Site conditions
- Commercial conditions

6. ELEMENTS OF PROJECT MANAGEMENT

The major elements of project management can be considered as :-

- Commitment
- Project planning
- Performance Measurement
- Man Management
- Activity Control
- Cost control
- Design concepts
- Logistics
- Monitoring
- Interfaces.

6.1 Commitment

Commitment is vital to the success of any project and the first step in gaining commitment is to fully involve all team members in the project. This can be relatively easily achieved in-house, but may be more difficult to accomplish in the customers organisation and the response to involvement depends a great deal on the clients own project management structure.

The customers team should be involved at all stages of the project but particularly :-

- Helping to plan the system, taking into account any preferred operating methods
- Seconding engineers onto the supplier's team for the specification phase
- Training system engineers, computer operators, maintenance engineers and line operators during installation and commissioning phases.

Trust between client and contractors is important in maintaining commitment. One way in which trust is nurtured is through good communications. This means a clear continuous flow of information between the parties, always channelled through the respective project managers. Other factors which help to build good relationships and trust are:-

- Responsiveness to requests
- Openness
- Frankness if problems do occur
- Timely reporting
- Complete information
- Competence

6.2 Project Planning

There are three key elements to project planning :-

- EXPERIENCE
- GOOD COMMUNICATION
- FREQUENT UPDATING AND REVIEW

Often, timescales established during sales phases are based on past experience of similar projects. Detailed planning is necessary to ensure that the project is completed in the time scale allowed.

Planning is a critical function and serious errors can have a devastating effect on a project. Personal Computer (P.C.) based systems can greatly assist with project planning. One example is :-

- "SUPER PROJECT PLUS"

This not only produces a critical path chart, but also has features such as resource level and resource cost analysis built in. To make the most of such systems they should be used for all current projects so that they can be linked to indicate the workload on shared resources (such as programmers, engineers etc.). If a resource is 'over loaded' the Project Manager is alerted and can take the appropriate action.

The project plan is evolved using the following basic rules:-

- List all activities
- Sequence activities
- Allocate resources
- Estimate time/cost for each activity
- Revise and refine plan

To make this task easier it is important to break a project down into small units. In determining the level of detail to which time elements should be monitored and controlled the Project Manager must strike a balance between having modules so small that he loses essential flexibility and spends all his time collecting data and so big that he cannot react quickly enough to emerging problems.

Such systems are especially helpful when a project has to be completed within an unusually short period of time.

One particular case is the installation carried out for Dixons Photographic Ltd. Dixons is one of the largest retailers of Hi-Fi, photographic and TV goods in the UK, with over 300 outlets, each of which has to be restocked weekly. The original warehouse comprised of 130,000 sq. ft. of conventional racking serviced by fork lift trucks. The decision was taken to automate and supplier choice was eventually based on who could guarantee the very short project duration required. One way of doing this was to phase the project and hand over parts of the system as soon as possible. Very close cooperation with the customer was essential, particularly as the system was also being installed in an existing building which remained in heavy use. However, it was more than anything else very tight project management which ensured that each phase went in on time.

Whilst it is very important to ensure that each event on a project plan is completed within the allotted time it is equally important to identify and make full use of opportunities to save time. Unknown and unexpected factors are always present in project engineering and modifications and improvisations will always have to be made to the most perfect of plans, so the underlying structure must be flexible enough to cope. Any accrued time savings may prove vital in this.

6.3 Performance Measurement

There are two main types of performance.

- Management performance in achieving project objectives
- System performance when installed and running.

Management performance is measured by the effectiveness of project execution, with regard to budget and timescale etc.

System performance is measured against the detail specification. In specifying system performance it is equally important to define how it is to be measured.

6.4 Man Management

It is most important that the Project Manager builds a dedicated team that he can rely on to work what can often in this Industry be very demanding hours. As such he should have considerable influence in choosing his team.

6.5 Activity Control

It is vital to keep tight control of all project activities. A Project Manager must be careful not to get too involved in detail and lose the overall perspective. On the other hand he has to be prepared to 'dive in' to help solve problems. To this end he must be given as much authority as possible to take decisions especially if working overseas.

6.6 Cost Control

Controlling costs is a vital aspect of project management.

It is part of the Project Manager's function to ensure that the quoted costs are not exceeded and additional discounts obtained if possible.

The mechanical effort to collect the necessary data for the Project Manager can also be reduced by use of computer packages.

6.7 Design Concepts

Standardisation

No two warehouses are identical and as such all projects can be said to be customised. However, systems do contain many elements which are similar; this applies to both hardware and software.

Considerable savings in both costs and time can be achieved by designing in appropriate levels of standardisation at module level. The reasons for this are primarily that sub-assemblies may then be produced in batches and repeat use of software modules greatly reduces both development and testing time, particularly on site.

Short term solutions

When designing systems it is not sufficient to just consider the end result. Of vital importance in many cases are the various intermediate problems which may occur, particularly if the system is to be introduced in phases for example in the case of a system refurbishment. With such projects it is often necessary to remove old equipment and install new equipment one subsystem at a time and this may require additional actions to be taken.

An example of this is a refurbishment project recently carried out on one of Nestle's distribution centres.

This system was first installed in 1970 and was controlled using a punch card system. The increasing workload in the warehouse and now inefficient order entry method meant that the control system had to be replaced. The project was planned in 2 stages. Initially the punch card system would be replaced by a DEC computer linked to the host IBM and in a second phase the existing crane controllers replaced with the latest microcomputers.

Requirements which had to be met during this process included:-
- No interruptions allowed in normal warehouse operations.
- No loss of data.
- Existing punch card system to be retained in an operational condition until the new system was thoroughly proven.

The project was tackled in the following way:-

- A microcomputer was used to act as a converter to process signals received from the existing equipment into signals suitable for the new DEC system. The program was fully tested using a simulation system.
- During closely defined time slots the 'converter' and controllers were linked to the system.
- Either the new or old systems could run with a switch over time of 2-3 minutes, which allowed engineers to test and commission the system without interruption to Nestle's operation.

In the second phase the crane controllers were installed at weekends whilst the warehouse was closed.

In most cases it is the Project Manager's responsibility to ensure that all such situations have been identified and reacted to and are correctly incorporated in his plans.

6.8 Logistical Problems

Not all problems associated with installation work actually occur on site. It often takes a major effort to get the equipment to site in the first place. An example where careful planning of the journey to site was required, concerned an installation for Danubia Petrochemie near Vienna in Austria.

Six thirty seven metre high crane masts each weighing 13 tonnes were fabricated in Switzerland and transported by rail to Vienna. Each mast required two railway carriages to support its full length. A careful check with both the Swiss and Austrian Railway Authorities meant that the train transporting the masts was scheduled in such a way that at no time was it diverted from the main line route and the actual rail route chosen so that no sharp curves were encountered. On site off loading and subsequent handling was achieved using a 300T capacity mobile crane with a 90M jib extension.

Once on site further problems may arise in getting large items of mechanical equipment into a building and once again these must be anticipated and planned for well in advance.

One such example of this occurred on a system supplied to the furniture retailer, IKEA. The problem was to lift the crane masts into the building from the side of the structure, not the end as is usual. This was again overcome using a mobile crane. Unfortunately the crane driver could not see the access hole in the roof and had to drive the crane "blind" relying upon directions relayed via a radio link from our Project Manager positioned on the roof of the building. The crane mast then had to be lowered through an existing aperture in the roof, with clearance of approx. 100mm on each side.

6.9 Project Monitoring

Projects should be monitored at two levels.

Although it is the Project Manager who is responsible for monitoring progress on a daily basis it is important that senior management review the project at a regular review meeting.

The review's purpose is to take an overview of each project and ensure that major requirements are being met and deviations to plan being acted on appropriately. In some cases the customer's senior management should also attend.

6.10 Interfaces

There are a wide range of interfaces which have to be defined and managed in a project which may involve several suppliers and the customers architects, planners etc.

Each interface has to be defined and the procedure for implementing it established.

The closer the interfaces the more accurately they must be defined and the more carefully they must be managed. This is particularly true with software but equally applies to interfaces between mechanical equipment from different suppliers or between mechanical equipment and the building.

An example of this is a system supplied to Zellweger of Switzerland. In this project the customer wanted to develop his own business management systems on the same computer that was to be used to control the automated warehouse. By very careful definition of the interfaces between the two systems this seemingly complicated task was carried out most successfully.

7. SUMMARY

To summarise, the following guide lines should help to maximise the chances of a project having a satisfactory outcome:-

- Appoint one person as project manager responsible for the entire project

- Use specialists for the project team

- Ensure clear and continuous communication between both team members and the client

- Develop a detailed specification at the beginning of a project. This should include details of the expected performance of the system and how the performance should be tested.

- Track the progress of a project and keep the plans up to date.

- Develop a clear project plan, which is detailed, but flexible enough to accept changes

- Evaluate all risks during the planning phase

- Develop a rapport with the customer

- Give particular attention to all interfaces.

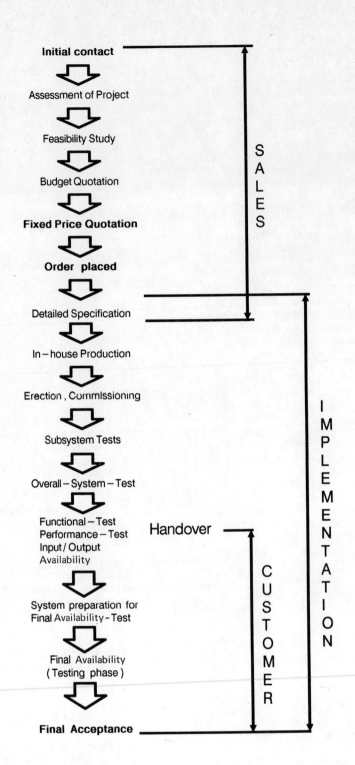

Fig 1 Project life cycle